Lecture Notes in Mathematics

Volume 2372

Editors-in-Chief

Jean-Michel Morel, City University of Hong Kong, Kowloon Tong, China
Bernard Teissier, IMJ-PRG, Paris, France

Series Editors

Karin Baur, University of Leeds, Leeds, UK
Michel Brion, UGA, Grenoble, France
Rupert Frank, LMU, Munich, Germany
Annette Huber, Albert Ludwig University, Freiburg, Germany
Davar Khoshnevisan, The University of Utah, Salt Lake City, UT, USA
Ioannis Kontoyiannis, University of Cambridge, Cambridge, UK
Angela Kunoth, University of Cologne, Cologne, Germany
Ariane Mézard, IMJ-PRG, Paris, France
Mark Podolskij, University of Luxembourg, Esch-sur-Alzette, Luxembourg
Mark Policott, Mathematics Institute, University of Warwick, Coventry, UK
László Székelyhidi, MPI for Mathematics in the Sciences, Leipzig, Germany
Gabriele Vezzosi, UniFI, Florence, Italy
Anna Wienhard, MPI for Mathematics in the Sciences, Leipzig, Germany

This series reports on new developments in all areas of mathematics and their applications - quickly, informally and at a high level. Mathematical texts analysing new developments in modelling and numerical simulation are welcome. The type of material considered for publication includes:

1. Research monographs
2. Lectures on a new field or presentations of a new angle in a classical field
3. Summer schools and intensive courses on topics of current research.

Texts which are out of print but still in demand may also be considered if they fall within these categories. The timeliness of a manuscript is sometimes more important than its form, which may be preliminary or tentative. Please visit the LNM Editorial Policy (https://drive.google.com/file/d/1MOg4TbwOSokRnFJ3ZR3ciEeKs9hOnNX_/view?usp=sharing)

Titles from this series are indexed by Scopus, Web of Science, Mathematical Reviews, and zbMATH.

Friedemann Schuricht • Moritz Schönherr

A Theory of Traces and the Divergence Theorem

Springer

Friedemann Schuricht
Faculty of Mathematics
TU Dresden
Dresden, Germany

Moritz Schönherr
Faculty of Mathematics
TU Dresden
Dresden, Germany

ISSN 0075-8434 ISSN 1617-9692 (electronic)
Lecture Notes in Mathematics
ISBN 978-3-031-86663-0 ISBN 978-3-031-86664-7 (eBook)
https://doi.org/10.1007/978-3-031-86664-7

This Springer imprint is published by the registered company Springer Nature Switzerland AG
The registered company address is: Gewerbestrasse 11, 6330 Cham, Switzerland

Dedicated with deep gratitude to

Eberhard Zeidler

Preface

Integration theory is an indispensable tool in modern analysis. It is based on a powerful measure theory where σ-additivity implies essential limit and continuity properties. Therefore, it is hard to imagine that an integration theory related to simply finitely additive measures could be significant. Largely unnoticed, already decades ago a comprehensive integration theory had been developed for such more general measures. However, one has to realize that central tools such as dominated convergence and the Radon-Nikodym theorem are not so strong. Therefore, that theory is not widely used and one finds few relevant applications.

During some of our previous investigations we were confronted with the dual of $\mathcal{L}^\infty(\Omega)$ for some $\Omega \subset \mathbb{R}^n$. It is well known that it coincides with $\mathcal{L}^1(\Omega)$ supplemented by certain finitely additive measures. These measures, however, are often considered unpleasant and not really useful. Nevertheless, we decided to take a closer look at them and found a nice book by Bhaskara Rao and Bhaskara Rao [4], which provides the state of the art on finitely additive measures available in the early 1980s. Somewhat hidden toward the end of the book, a very interesting characterization particularly attracted our attention (cf. [4, p. 244]).

To illustrate, let us consider an open set $\Omega \subset \mathbb{R}^n$. Then the dual of $\mathcal{L}^\infty(\Omega)$ can be identified with all finitely additive measures μ on Ω that vanish on sets A with zero Lebesgue measure $\mathcal{L}^n(A)$. This includes both measures that are σ-additive (these correspond to $\mathcal{L}^1(\Omega)$) and others that are not. Now this interesting characterization says that μ is purely finitely additive if and only if there is some decreasing sequence $A_k \subset \Omega$ such that $|\mu|(\Omega \setminus A_k) = 0$ and $\mathcal{L}^n(A_k)$ tends to zero. This means that each A_k contains the total mass of μ, but $|\mu|(\bigcap_k A_k) = 0$ by $\mathcal{L}^n(\bigcap_k A_k) = 0$. Certainly such a measure μ, if nontrivial, cannot be σ-additive. But do such strange measures actually exist? We found only very few examples and most of them seemed to be rather theoretical and not really relevant to applications. We have the impression that, until now, not much attention has been paid to this special property due to this apparent lack of applications.

Quite soon the density $\operatorname{dens}_x A$ of a set A at point $x \in \Omega$ came to our minds. This has long been an important tool in geometric measure theory. However, it seems that previously it had not really been considered as set function. But, in doing so, it becomes immediately clear that $\operatorname{dens}_x A$ is finitely additive and cannot be σ-additive. Since the sets for which $\operatorname{dens}_x A$ is defined do not form an algebra, some suitable extension is required in order to obtain a finitely additive measure. Though not unique, the Hahn-Banach theorem ensures such an extension on the Borel sets. Clearly the obtained extension, again called dens_x, vanishes on Lebesgue null sets and thus belongs to the dual of $\mathcal{L}^\infty(\Omega)$. To our surprise we were not able to find this quite simple example in the literature. Obviously dens_x has some similarity with the Dirac measure δ_x concentrated at x. But notice that $\operatorname{dens}_x(\{x\}) = 0$ while $\delta_x(\{x\}) = 1$. The characterizing property for purely finitely additive measures mentioned above is obviously satisfied by $A_k = B_{1/k}(x)$, where $B_r(x)$ is the ball around x with radius r. What does that mean for the related integral? We must have

$$\int_\Omega f \, d \operatorname{dens}_x = \int_{A_k} f \, d \operatorname{dens}_x \quad \text{for all} \quad A_k \, .$$

Hence the integral has the strange property that it solely depends on the values of f on any small neighborhood of x and, moreover, f need not to be uniquely defined at x. Can something like that be useful? We realize that the precise representative $\tilde{f}(x)$ of an integrable function f at x, that is the limit of some averages and exists for $\mathcal{L}^n$-a.e. $x \in \Omega$, has some similar feature. It turns out that

$$\tilde{f}(x) = \int_\Omega f \, d \operatorname{dens}_x \quad \text{for } \mathcal{L}^n\text{-a.e. } x \, .$$

In other words we obtain some integral representation for the precise representative.

Such a limit of averages is also encountered in the pointwise computation of the trace function $f^{\partial\Omega}$ of a Sobolev function f on a suitable Ω. This is needed in the fundamental Gauss-Green formula

$$\int_\Omega f \operatorname{div} \varphi \, d\mathcal{L}^n + \int_\Omega \varphi \cdot Df \, d\mathcal{L}^n = \int_{\partial\Omega} f^{\partial\Omega} \varphi \cdot \nu^{\partial\Omega} \, d\mathcal{H}^{n-1}$$

where φ is typically a test function in $C^1(\overline{\Omega}, \mathbb{R}^n)$. The necessity of a trace $f^{\partial\Omega}$ and of an outer normal field $\nu^{\partial\Omega}$ on $\partial\Omega$ leads to essential limitations of this important formula. Can finitely additive measures be useful here? Using constructions similar to those for $\operatorname{dens}_x A$, we get purely finitely additive measures ω on Ω that "live" on any small neighborhood of the boundary $\partial\Omega$. For the condition discussed above we can take decreasing neighborhoods A_k of $\partial\Omega$ within Ω and we even have $\bigcap_k A_k = \emptyset$. It can be shown that there is some finitely additive measure ω and some normal field ν on Ω such that

$$\int_{\partial\Omega} f^{\partial\Omega} \varphi \cdot \nu^{\partial\Omega} \, d\mathcal{H}^{n-1} = \int_\Omega f \varphi \cdot \nu \, d\omega \, .$$

Notice that, on the right hand side, we have a function f and a normal field ν that are only defined on Ω but not on $\partial\Omega$. Inherently the integral on the right carries out the limit procedure of averages near $\partial\Omega$ as done for $f^{\partial\Omega}$. This suggests that finitely additive measures are closely related to traces and it appears that they might open up a completely new perspective for a general theory of traces. In this book we will develop a foundation of such a theory. It turns out that the presented approach applies easily to more general sets Ω, to inner boundaries, and to singularities that had not been tractable before.

At first glance the integration theory based on finitely additive measures appears to be less useful since it lacks the familiar strong continuity properties. However, a scarcely noticed interesting averaging property, that is not accessible through the classical theory, turns out to be a very powerful and natural tool for the treatment of traces. This, in particular, shows that finitely additive measures and the related integrals develop their potential using features totally different from those of classical theory.

Dresden, Germany
September 2024

Friedemann Schuricht
Moritz Schönherr

Acknowledgment

We wish to express our deep gratitude to Eberhard Zeidler and Stuart Antman for their inestimable support and the profound scientific stimulation to the first author.

! Warning

Some of our terminology deviates from standard usage. In particular notice:

- The notion *measure* is used for any finitely additive measure while σ-*measure* is used for σ-additive measures.
- Section 2.1 summarizes preliminary material about measures from [4]. Here we changed terminology to the one commonly used in measure theory and, in doing so, we deviate substantially from [4].

Contents

Chapter 1
Introduction

The divergence theorem belongs to the most important tools in mathematical analysis and continuum physics and goes back to Gauss (1813), Ostrogradskii (1826), and Green (1828) (cf. Stolze [45] for a brief history). It connects a volume integral with an integral over the bounding surface. For a regular bounded open set $\Omega \subset \mathbb{R}^n$ and a regular vector field $F : \Omega \to \mathbb{R}^n$ it says that

$$\int_{\Omega} \operatorname{div} F \, d\mathcal{L}^n = \int_{\partial\Omega} F \cdot \nu^{\Omega} \, d\mathcal{H}^{n-1} \tag{1.1}$$

where $\mathcal{L}^n$ is the Lebesgue measure, $\mathcal{H}^{n-1}$ the $(n-1)$-dimensional Hausdorff measure, and ν^{Ω} the outer unit normal to Ω. In physics it combines volume and surface phenomena and is indispensable for fundamental balance and conservation laws. If we apply (1.1) to a product fF with some regular $f : \Omega \to \mathbb{R}$, then we obtain the Gauss-Green formula, also called integration by parts formula,

$$\int_{\Omega} \big(f \operatorname{div} F + F \cdot Df\big) \, d\mathcal{L}^n = \int_{\partial\Omega} fF \cdot \nu^{\Omega} \, d\mathcal{H}^{n-1} \,. \tag{1.2}$$

This formula is the basis for the definition of weak derivatives and, according to Eberhard Zeidler, it is "*the key to the modern theory of partial differential equations and to modern calculus of variations*" (cf. [54, p. 119]). It is of course desirable to have such an essential formula available for a very large class of sets Ω and functions f and F. But it turns out that the surface integral leads to crucial limitations. On the one hand we need sets Ω having some normal field on their boundaries and on the other hand we need functions F, f on Ω that possess a pointwise trace function on $\partial\Omega$. In practice we need a balance in regularity for these ingredients. This leads to the typical cases where, roughly speaking, either F, f have to be smooth if Ω has merely finite perimeter (that are the most general sets with some kind of normal field on the boundary) or we can allow that fF has merely weak derivatives while Ω has to have Lipschitz boundary. In continuum mechanics this situation prevents that we

F. Schuricht, M. Schönherr, *A Theory of Traces and the Divergence Theorem*,
Lecture Notes in Mathematics 2372, https://doi.org/10.1007/978-3-031-86664-7_1

can use (1.2) on sets Ω where concentrations occur on $\partial\Omega$ or on sets Ω not having a normal field on $\partial\Omega$. More precisely, we are unable to compute the flux through the boundary $\partial\Omega$ for such sets. However, nature "knows" what happens or, with the words of Albert Einstein (cf. [26]): *"God does not care about our mathematical difficulties; He integrates empirically."* Therefore the derivation of more general versions of (1.2) is an important task. But, without intending completeness, let us first sketch some developments going beyond smoothness.

We start with the simple observation that the regularity of the product fF is essential for the availability of (1.2). This means that a weak regularity of one factor requires a stronger regularity of the other factor. But also a bad boundary $\partial\Omega$ requires a more regular product fF than a good one and it turns out that a Lipschitz boundary is apparently always a good boundary. Moreover, in cases where Ω is not open, one needs some ambient open set U containing Ω such that f and F with their derivatives can be reasonably defined on U.

In the theory of partial differential equations the Gauss-Green formula is closely related to the treatment of boundary value problems where, during the last century, Sobolev functions $f \in \mathcal{W}^{k,p}(\Omega)$ became more and more important. But these functions are merely defined $\mathcal{L}^n$-a.e. on Ω and their restriction to $\partial\Omega$ does not make sense. Thus the derivation of a trace operator

$$T : \mathcal{W}^{k,p}(\Omega) \to \mathcal{L}^1(\partial\Omega, \mathcal{H}^{n-1}),$$

that assigns reasonable values on the boundary, is an essential requirement. Since Sobolev functions are not just integrable but also have an integrable weak derivative, the limit of averages near the boundary exists $\mathcal{H}^{n-1}$-a.e. on $\partial\Omega$ and provides a trace operator if e.g. Ω is a bounded open set with Lipschitz boundary. It turns out that this is also true for functions of bounded variation (BV functions) $f \in \mathcal{BV}(\Omega)$, that played an increasing role during the last decades. This way one obtains (1.2) for Sobolev and BV functions f, for Lipschitz continuous vector fields $F \in \mathrm{Lip}(\overline{\Omega})$, and for open bounded Ω with Lipschitz boundary, where the second term on the left hand side becomes $\int_\Omega F\, dDf$ for BV functions (cf. [32, p. 168]).

The balance in regularity for the product fF is nicely worked out by Anzellotti [3] for bounded open Ω with Lipschitz boundary. He treats the pairing of certain BV functions f with bounded vector fields $F \in \mathcal{L}^\infty(\Omega)$ where the distributional divergence $\operatorname{div} F$ either belongs to $\mathcal{L}^p(\Omega)$ or it is just a Radon measure. In this generality the product $F \cdot Df$ in (1.2) has to be interpreted as a measure. Moreover one cannot assign a pointwise trace function to F on $\partial\Omega$, but merely the normal component $F \cdot \nu^\Omega$ has a so-called normal trace $\mathcal{H}^{n-1}$-a.e. on $\partial\Omega$ (cf. also Kawohl-Schuricht [27, p. 540] for a variant that is relevant for the treatment of the 1-Laplace operator).

The development of geometric measure theory lead to a different substantial improvement, since it allowed the extension of (1.2) to the large class of Ω having merely finite perimeter. Such sets have a measure theoretic outer unit normal ν^Ω for $\mathcal{H}^{n-1}$-a.e. point on the measure theoretic boundary $\partial_*\Omega$ (that is contained in the topological boundary $\partial\Omega$) and (1.2) becomes

$$\int_{\Omega} \big(f \operatorname{div} F + F \cdot Df\big)\, d\mathcal{L}^n = \int_{\partial_* \Omega} f F \cdot \nu^{\Omega}\, d\mathcal{H}^{n-1}\,. \tag{1.3}$$

As price for the weak regularity of the set Ω, one had to require initially that fF belongs to $\mathrm{Lip}_c(\mathbb{R}^n, \mathbb{R}^n)$, i.e. Lipschitz continuous with compact support (cf. De Giorgi [19], Federer [23, 24]). These results have then been extended to bounded vector fields of bounded variation, i.e. belonging to $\mathcal{BV}(\Omega, \mathbb{R}^n)$ (cf. Vol'pert [47, 48]). Pairings of bounded F where $\operatorname{div} F$ is a Radon measure with bounded BV functions f on sets of finite perimeter are treated by Crasta-De Cicco [15, 16] and Crasta-De Cicco-Malusa [17].

Later on it became more and more common to consider the left hand side in (1.2) as linear functional in f or in F. This is an obvious idea if one has in mind the potential of linear functionals in modern analysis (duality, weak derivatives etc.). Let us now focus on the case where F is a general vector field with weak regularity as needed in applications and let f play the role of a more regular test function, a situation that is mostly met in the literature (it is a simple task to transfer the results to the opposite case). It turned out that, with $U \subset \mathbb{R}^n$ open, with $\operatorname{div} F$ taken in the distributional sense and with $1 \le p \le \infty$,

$$\mathcal{DM}^p(U) = \{F \in \mathcal{L}^p(U, \mathbb{R}^n) \mid \operatorname{div} F \text{ is a Radon measure}\}$$

is a reasonable space for the selection of the vector field F. Thus we consider the left hand side in (1.2), adapted to such general F and for some $\Omega \subset U$, as linear functional

$$T_F(f) = \int_{\Omega} f\, d \operatorname{div} F + \int_{\Omega} F \cdot Df\, d\mathcal{L}^n \quad \text{for} \quad f \in X \tag{1.4}$$

where we basically find the following choices for X in the literature

$$\mathrm{Lip}(U)\,, \;\; \mathrm{Lip}_{\mathrm{loc}}(U)\,, \;\; C^1(U)\,, \;\; \{f \in C(U) \mid Df \in \mathcal{L}^{p'}(U)\}\,.$$

All choices have in common that, for suitable $\Omega \subset U$, we somehow have that $X \subset C(\overline{\Omega})$. Moreover it is an important observation that $T_F(f)$ in fact merely depends on the values of f on $\partial\Omega$ (cf. Šilhavý [41, p. 22], [43, p. 448], Chen-Frid [6, p. 262], Chen-Torres-Ziemer [10, p. 254]). Therefore most investigations are devoted to the natural question for which F the functional T_F is related to a Radon measure on $\partial\Omega$ or on $\partial_*\Omega$.

The strongest results are obtained for $F \in \mathcal{DM}^\infty(U)$ and for sets of finite perimeter $\Omega \Subset U$ (i.e. compactly contained in U). Here also the measure theoretic interior $\mathrm{int}_*\Omega$ is used (that differs from Ω merely by an $\mathcal{L}^n$-null set). For such F one generally has that $\operatorname{div} F$ is absolutely continuous with respect to $\mathcal{H}^{n-1}$ (cf. Šilhavý [41, p. 21], Chen-Torres [8, p. 250]). This implies that T_F is always a Radon measure on $\partial_*\Omega$. If $(\operatorname{div} F)(\partial_*\Omega) = 0$ and if F is a precise representative such that $\mathcal{H}^{n-1}$-a.e. $x \in \partial_*\Omega$ is a Lebesgue point, then one has the slightly modified version of (1.3) that

$$\int_{\operatorname{int}_*\Omega} f \, d \operatorname{div} F + \int_\Omega F \cdot Df \, d\mathcal{L}^n = \int_{\partial_*\Omega} f F \cdot \nu^\Omega \, d\mathcal{H}^{n-1} \tag{1.5}$$

(cf. Degiovanni [20, p. 212]). One also has (1.5) in the case of continuous F (cf. Šilhavý [42, p. 85], Chen-Torres-Ziemer [10, p. 284], Comi-Payne [13, p. 198]). Otherwise one has to use some approximation to get a normal trace function $t^{\text{int}} \in \mathcal{L}^\infty(\partial_*\Omega, \mathcal{H}^{n-1})$ such that

$$\int_{\operatorname{int}_*\Omega} f \, d \operatorname{div} F + \int_\Omega F \cdot Df \, d\mathcal{L}^n = \int_{\partial_*\Omega} f t^{\text{int}} \, d\mathcal{H}^{n-1} \tag{1.6}$$

(cf. Šhilhaý [41, p. 25] where (1.3) is applied to smooth approximations of F and Chen-Frid [5, p. 97–98], Chen-Torres [8, p. 252] where approximations of Ω are used to obtain t^{int} as weak* limit of Radon measures; cf. also Comi-Payne [13, pp. 194, 200], Comi-Torres [14], Šilhavý [44, p. 6]). Since F can have jumps across the boundary such that $(\operatorname{div} F)(\partial_*\Omega) \neq 0$, it is reasonable to apply the previous result to the measure theoretic exterior $\operatorname{ext}_*\Omega$ of Ω. This readily gives some $t^{\text{ext}} \in \mathcal{L}^\infty(\partial_*\Omega, \mathcal{H}^{n-1})$, that differs from t^{int} in general, such that

$$\int_{\operatorname{int}_*\Omega \cup \partial_*\Omega} f \, d \operatorname{div} F + \int_\Omega F \cdot Df \, d\mathcal{L}^n = \int_{\partial_*\Omega} f t^{\text{ext}} \, d\mathcal{H}^{n-1}$$

(cf. Chen-Torres-Ziemer [10, pp. 275, 281], Comi-Payne [13, pp. 194, 200]). The extension $\hat{F}$ of $F \in \mathcal{DM}^\infty(U)$ by zero belongs to $\mathcal{DM}^\infty(\mathbb{R}^n)$ if U is open and bounded and satisfies $\mathcal{H}^{n-1}(\partial\Omega \setminus \operatorname{ext}_*\Omega) = 0$ (cf. Chen-Li-Torres [12, p. 242]; cf. also Chen-Torres [8, p. 258], Chen-Torres-Ziemer [10, p. 288], Comi-Payne [13, p. 209]). In this case one can readily get the previous results for Ω with finite perimeter that do not need to be compactly contained in U by applying the previous results to $\hat{F}$ on some larger U (cf. Chen-Li-Torres [12, p. 248]). In applications to shock waves or cracks it is also desirable to have a Gauss-Green formula with boundary integral on the complete topological boundary, i.e. where the inner part of the boundary is also taken into account. This can easily be derived from (1.6) by moving the part of the first integral on the left hand side over $\partial\Omega \cap \operatorname{int}_*\Omega$ to the right hand side. If Ω is open, then solely the integral over Ω remains on the left hand side (cf. Chen-Li-Torres [12, p. 248] and Remark 4.36 below). Let us also refer to Leonardi-Saracco [30] for some special result in $\mathbb{R}^2$.

For general vector fields $F \in \mathcal{DM}^1(U)$ and sets $\Omega \Subset U$ with finite perimeter we have (1.5) if $(\operatorname{div} F)(\partial_*\Omega) = 0$ and if F is a precise representative such that both $\mathcal{H}^{n-1}$-a.e. $x \in \partial_*\Omega$ is a Lebesgue point and F is $\mathcal{H}^{n-1}$-integrable on $\partial_*\Omega$ (cf. Degiovanni [20, p. 212], Šilhavý [40, p. 237]). We also have (1.5) if the vector field F is continuous (cf. Šilhavý [42, p. 85], Chen-Torres-Ziemer [10, p. 284],

Comi-Payne [13, p. 198]; cf. also Chen-Comi-Torres [11, p. 131]). For general open and closed sets Ω there are merely some approximation results as stated in Proposition 4.2 below (cf. Schuricht [38, p. 534], Schuricht [39, p. 189], Šilhavý [43, p. 449], Chen-Comi-Torres [11, pp. 117–123]). It turns out, however, that even for Ω with finite perimeter, T_F needs not to be a Radon measure on the boundary of Ω (cf. Šilhavý [43, p. 449] for an example). In the general case we find several sufficient conditions for T_F to be a Radon measure on the boundary (cf. Šilhavý [43, p. 449], Šilhavý [42, p. 84], Chen-Comi-Torres [11, pp. 127–129]). Conditions for the existence of an integrable density are given in Šilhavý [41, p. 26]. Let us also mention that many results of Šilhavý are worked out for the more general case where $F \in \mathcal{DM}(U)$, i.e. a vector-valued Radon measure such that the distributional divergence is also a Radon measure. We do not consider this generality in the present treatment. But, due to the product rule that φF also belongs to $\mathcal{DM}(U)$ for any $\varphi \in \mathcal{W}^{1,\infty}(U)$, it might be possible to extend essential results to that generality (cf. Šilhavý [43, p. 448]).

Let us come back to the linear functional T_F and the underlying space X. As already mentioned, in previous investigations X was somehow always contained in $C(\overline{\Omega})$. Thus it was natural to ask to what extent T_F is continuous on $C(\overline{\Omega})$ and, if this is the case, if it is representable by a Radon measure on $\partial\Omega$ as an element of the dual space. Since $T_F(f)$ merely depends on $f_{|\partial\Omega}$, this seems to be an obvious strategy. The drawback is that not all T_F can be represented by a Radon measure on $\partial\Omega$ and that there are rarely results beyond Ω with finite perimeter.

Therefore let us discuss a different choice for X. If one looks at the right hand side in (1.4), the optimal pairing of $F \in \mathcal{DM}^1(U)$ seems to be with $f \in \mathcal{W}^{1,\infty}(U)$. In this case one trivially gets that T_F is a continuous linear functional on $X = \mathcal{W}^{1,\infty}(U)$. Therefore a representation of T_F as dual element of $\mathcal{W}^{1,\infty}(U)$ is possible for all $F \in \mathcal{DM}^1(U)$ and all Borel sets $\Omega \subset U$. But the question is how far this is useful. The dual of $\mathcal{W}^{1,\infty}(U)$ can be identified with a product of dual spaces $\mathcal{L}^\infty(U)^*$ and it is known that $\mathcal{L}^\infty(U)^*$ consists of $\mathcal{L}^1(U)$ supplemented by certain finitely additive measures. Since the integration theory for finitely additive measures is commonly considered as not very powerful, one could hope that the functionals T_F are in fact related to measures with an $\mathcal{L}^1$-integrable density. Notice that, in the simple case of a smooth F, we can trivially identify T_F with $(\operatorname{div} F, F) \in \mathcal{L}^1(U)^{n+1}$ as an element of the dual of $\mathcal{W}^{1,\infty}(U)$ (cf. (1.4)). But this is not what we want to get and it gives no improvement. We are rather looking for some representation on or at least near the boundary of Ω. Therefore let us choose $\delta > 0$, let $\chi_\delta \in C_c^1(U)$ be supported on the δ-neighborhood $(\partial\Omega)_\delta$ of $\partial\Omega$ and let $\chi_\delta = 1$ on $(\partial\Omega)_{\delta/2}$. Then it is a standard result that

$$T_F(f) = T_F(\chi_\delta f) = T_{\chi_\delta F}(f) \quad \text{for all } f\,.$$

This motivates to replace T_F with the functional $T_F^\delta = T_{\chi_\delta F}$ that merely considers values of f near $\partial\Omega$. This way we get that T_F can be localized near $\partial\Omega$ and, similar as above, we can identify T_F^δ with $(\operatorname{div}(\chi_\delta F), \chi_\delta F) \in \mathcal{L}^1(U)^{n+1}$ as dual element of $\mathcal{W}^{1,\infty}(U)$. But such a representation always depends on $\delta > 0$ and it turns out that we cannot remove that dependence in general with dual elements belonging to $\mathcal{L}^1(U)$.

Hence let us briefly set aside our reservations about finitely additive measures and take a brief look at them. In continuum mechanics it is a simple observation that contact interactions are naturally related to finite additivity (cf. Schuricht [38, p. 512]). Even more, it seems that finite additivity is characterizing for short-range phenomena (cf. Schuricht [39]). Let us illustrate this by a simple example. The density of a set $A \subset \mathbb{R}^n$ at point $x \in \mathbb{R}^n$ is an important standard tool in geometric measure theory and it is given by

$$\mathrm{dens}_x A = \lim_{\delta \downarrow 0} \frac{\mathcal{L}^n(A \cap B_\delta(x))}{\mathcal{L}^n(B_\delta(x))}$$

whenever the limit exists and where $B_\delta(x)$ is the open ball with radius $\delta > 0$ centered at x. Let us fix x and consider $A \to \mathrm{dens}_x A$ as set function. We readily see that it is additive for disjoint sets. Now we fix some $y \neq x$ and a sequence $B_k = B_{\delta_k}(y)$ of open balls with $\delta_k \uparrow |y - x|$. Then

$$\mathrm{dens}_x B_k = 0 \quad \text{for all } k, \quad \text{but} \quad \mathrm{dens}_x \Big(\bigcup_{k \in \mathbb{N}} B_k \Big) = \frac{1}{2},$$

which obviously prevents σ-additivity. We readily observe that we merely have to know the intersection of A with an arbitrarily small neighborhood of x for the determination of $\mathrm{dens}_x A$. But notice that $\mathrm{dens}_x(\{x\}) = 0$. Hence, roughly speaking, dens_x lives near x but it is not supported at the point x. By a simple Hahn-Banach argument dens_x can be extended, though not uniquely, to a finitely additive measure on all Borel sets. It turns out that this example is typical for a certain class of finitely additive measures μ. They can be characterized by the fact that there is a decreasing sequence of sets A_k such that

$$\mathcal{L}^n(A_k) \to 0 \quad \text{and} \quad |\mu|(A_k^c) = 0 \ \text{ for all } \ k \in \mathbb{N}$$

where $|\mu|$ is the total variation of μ and A_k^c denotes the complement of A_k. But to what extend can that be useful for traces of functions? First we readily observe some similarity of traces to the measure dens_x since the computation of the trace $\tilde{f}(x)$ of e.g. a Sobolev function f requires the values of f merely in an arbitrarily small neighborhood of x. For a more precise look we observe that one can define an integral for the finitely additive measure dens_x similar to the usual integral for σ-additive measures by, roughly speaking, just replacing "convergence a.e." with "convergence in measure" in the definition. Then, for any $f \in L^1(\Omega)$ with Ω open and for any Lebesgue point $x \in \Omega$, we obtain that f is dens_x-integrable with

$$\int_\Omega f \, d\,\mathrm{dens}_x = f(x)\,.$$

Moreover, let $f \in \mathcal{W}^{1,1}(\Omega)$ be a Sobolev function on some open Ω with Lipschitz boundary and let dens_x^Ω be a measure similar to dens_x but, for $x \in \partial\Omega$, giving the density within Ω (cf. (2.13) below). Then we have for a trace function $\tilde{f} \in \mathcal{L}^1(\partial\Omega, \mathcal{H}^{n-1})$ of f that

$$\int_\Omega f \, d\,\mathrm{dens}_x = \tilde{f}(x) \quad \text{for } \mathcal{H}^{n-1}\text{-a.e. } x \in \partial\Omega$$

(cf. Proposition 2.15 below). Therefore we can compute the trace $\tilde{f}(x)$ by an integral over Ω instead of the usual limit of mean values. But this means that integrals for finitely additive measures not only provide an integral calculus for traces but also for the precise representative of an $\mathcal{L}^1$-function. These observations suggest that finitely additive measures are a very convenient and natural tool for the treatment of traces.

In this book we thus develop a general theory of traces that relies on the dual of $\mathcal{L}^\infty(U)$. More precisely, we understand a trace T on some set $\Gamma \subset U$ as a linear continuous functional T in φ that depends only on the values of φ near Γ and we focus on the special case of functionals on $\mathcal{L}^\infty(U)$ or $\mathcal{W}^{1,\infty}(U)$. It turns out that T_F from (1.4) is a trace on $\partial\Omega$ and, this way, we derive Gauss-Green formulas for any $F \in \mathcal{DM}^1(U)$ on any Borel set $\Omega \subset U$. In the most general case we get for each $\delta > 0$ the existence of a scalar measure $\lambda_F \in \mathcal{L}^\infty(U)^*$ and a vector measure $\mu_F \in \mathcal{L}^\infty(U, \mathbb{R}^n)^*$ such that

$$\langle T_F, \varphi \rangle := \int_\Omega \varphi \, d \operatorname{div} F + \int_\Omega F \cdot D\varphi \, d\mathcal{L}^n = \int_{(\partial\Omega)_\delta} \varphi \, d\lambda_F + \int_{(\partial\Omega)_\delta} D\varphi \, d\mu_F$$

for all $\varphi \in \mathcal{W}^{1,\infty}(U)$ (cf. Proposition 4.1). If the functional T_F is finite in some sense (cf. (4.5)), then we can remove the dependence on δ and we write

$$\langle T_F, \varphi \rangle = \int_\Omega \varphi \, d \operatorname{div} F + \int_\Omega F \cdot D\varphi \, d\mathcal{L}^n = \fint_{\partial\Omega} \varphi \, d\lambda_F + \fint_{\partial\Omega} D\varphi \, d\mu_F \tag{1.7}$$

where $\fint_{\partial\Omega}$ means that we have to integrate over any small neighborhood of $\partial\Omega$. In this case we typically have that the measures λ_F, μ_F are merely finitely additive. This Gauss-Green formula precisely accounts for boundary points belonging to Ω, which is important in the case of concentrations of $\operatorname{div} F$ on $\partial\Omega$. Thus, in addition to the usually considered open and closed Ω, we can exactly treat any Borel set Ω' with $\operatorname{int}\Omega \subset \Omega' \subset \overline{\Omega}$. We also show that the second term on the right hand side cannot be neglected in general (e.g. if T_F restricted to $\varphi \in C(\overline{\Omega})$ does not correspond to a Radon measure). If T_F can be extended to a linear continuous functional on

$\mathcal{L}^\infty(U)$ (i.e. if T_F is continuous with respect to the $\mathcal{L}^\infty$-norm), then we can choose $\mu_F = 0$ (cf. Proposition 4.5). However the choice of λ_F, μ_F is not unique in general. We provide examples where $\mu_F = 0$ in (1.7) is possible but, alternatively, also $\lambda_F = 0$ with some nontrivial μ_F can be chosen. If $\mu_F = 0$ and $\Omega \Subset U$, then all $\varphi \in \mathcal{W}^{1,\infty}(U)$ belong to $C(\overline{\Omega})$ and we can identify λ_F with a Radon measure supported on $\partial\Omega$. This way we somehow recover previous results. But notice that we typically have to integrate over Ω near $\partial\Omega$ in (1.7) and not over $\partial\Omega$. This raises the question how far a Radon measure or a trace function on $\partial\Omega$ is really needed.

Though the measures λ_F and μ_F are not unique in general, they have to be linear in F as a whole and, of course, some explicit dependence of the "boundary integrals" on F would be desirable. This turns out to be possible for Ω with finite perimeter if we can choose $\mu_F = 0$ and if F is appropriate (e.g. essentially bounded). However we do not intend to get a Radon measure on the measure theoretic boundary in this case. Therefore we replace the commonly used pointwise normal field ν^Ω by some finitely additive normal measure ν^Ω that might be constructed e.g. by means of the signed distance function of $\partial\Omega$. This way we provide several choices for ν^Ω, e.g. related to an open Ω or to a closed Ω. If F is ν^Ω-integrable and satisfies some compatibility condition (cf. (4.56)), then we have that

$$\langle T_F, \varphi\rangle = \int_\Omega \varphi\, d\,\mathrm{div}\, F + \int_\Omega F \cdot D\varphi\, d\mathcal{L}^n = \fint_{\partial\Omega} \varphi F\, d\nu^\Omega\,. \tag{1.8}$$

If Ω is open, then ν^Ω only uses values of φF near $\partial\Omega$ inside Ω and if it is closed, then only values outside of Ω are used. In (1.8) one can even include a certain weight function χ for div F on $\partial\Omega$ (cf. (4.60)). Since ν^Ω belongs to $\mathcal{L}^\infty(U, \mathbb{R}^n)^*$, all essentially bounded $F \in \mathcal{DM}^\infty(U)$ are trivially ν^Ω-integrable and they also satisfy the additional compatibility condition. In addition, (1.8) is also applicable to certain unbounded vector fields F. For normal measures based on the signed distance function of $\partial\Omega$ we get even more structural information for the boundary term. If e.g. Ω is open and satisfies some mild perimeter bound (cf. (4.49)), we can use a suitable normal measure to derive the more explicit form

$$\langle T_F, \varphi\rangle = \int_\Omega \varphi\, d\,\mathrm{div}\, F + \int_\Omega F \cdot D\varphi\, d\mathcal{L}^n = \fint_{\partial\Omega} \varphi F \cdot \nu^\Omega\, d\omega_{\partial\Omega}^{\mathrm{int}}$$

where ν^Ω is the normal field on Ω given by the gradient of the signed distance function for $\partial\Omega$ and $\omega_{\partial\Omega}^{\mathrm{int}}$ is, roughly speaking, a finitely additive extension of the Radon measure $\mathcal{H}^{n-1}\lfloor\partial\Omega$. This version of the Gauss-Green formula is quite close to the usual form (1.3) with an integral on $\partial_*\Omega$. The advantage here is that we do not need an explicit trace function of F on $\partial\Omega$, since the knowledge of F for $\mathcal{L}^n$-a.e. point on a neighborhood of $\partial\Omega$ is sufficient to compute the integral. Moreover, in contrast to the measure theoretic boundary, the formula uses the topological boundary $\partial\Omega$ that takes into account also inner parts of the boundary. This is desirable for the treatment of cracks and shocks. Notice that in the case where $\Omega = U$ is open, the functions in $\mathcal{W}^{1,\infty}(U)$ belong to $C(\Omega)$ but they do not need to be in $C(\overline{\Omega})$. Therefore it is possible to change φ independently on both sides of some inner crack or shock which allows a precise description of the situation on each side separately.

The results can be easily transferred to Gauss-Green formulas for Sobolev and BV functions f with test functions $\varphi \in \mathcal{W}^{1,\infty}(U, \mathbb{R}^n)$. For open Ω with Lipschitz boundary we supplement the classical Gauss-Green formula (1.2) with

$$\int_\Omega \big(f \operatorname{div} \varphi + \varphi \cdot Df\big)\, d\mathcal{L}^n = \fint_{\partial\Omega} f\varphi \, d\nu^{\mathrm{int}} = \fint_{\partial\Omega} f\varphi \cdot \nu^\Omega \, d\omega^{\mathrm{int}}_{\partial\Omega} \tag{1.9}$$

where ν^{int} is some normal measure and $\omega^{\mathrm{int}}_{\partial\Omega}$ is a measure as above. Notice that, similar as for vector fields, we do not need an explicit trace function for f on the boundary. One could say that its computation is somehow incorporated into the integral. As application of the trace theory we finally show for some boundary value problem containing the p-Laplace operator that there exists a weak solution for any bounded open set Ω.

Let us now briefly sketch how this book is organized. In Chap. 2 we start with some rough introduction to the integration theory for finitely additive measures, since it is not so well known. To this end we summarize material that is widely scattered throughout the book of Bhaskara Rao and Bhaskara Rao [4] and that cannot be found elsewhere in such a compact form. But notice that the given survey is far from providing all the results on integration theory which are required for our subsequent investigation. This material is supplemented by some typical examples and by some new results that are used later. To avoid confusion let us add the following warning: *We use the notion measure for any finitely additive measure while σ-additivity is indicated by the notion σ-measure. Moreover we orient our terminology to the one commonly used in measure theory and, in doing so, we substantially deviate from the terminology used in the underlying book [4].*

Chapter 3 presents a general approach to traces on arbitrary sets. In Sect. 3.1 we first define traces as certain linear continuous functionals. Then we provide a simple but important class of trace functionals over $\mathcal{L}^\infty(U)$. They are needed for later use, but initially they also serve for illustration. In Sect. 3.2 we show that T_F from (1.4) is a trace functional on $\partial\Omega$ over $\mathcal{W}^{1,\infty}(U)$ for any $F \in \mathcal{DM}^1(U)$ and any Borel set $\Omega \subset U$. We also get an analogous result for Sobolev functions in $\mathcal{W}^{1,1}(U)$ and for BV functions in $\mathcal{BV}(U)$. Section 3.3 is devoted to the representation of such traces by means of measures that are "living" near $\partial\Omega$. Here we distinguish three variants of generality called (G), (L), and (C). Theorem 3.14 and several necessary and sufficient conditions for certain special cases are the basis for the subsequent Gauss-Green formulas. Some examples illustrate the spirit behind the three variants.

Chapter 4 presents several divergence theorems or, more generally, several Gauss-Green formulas. In Sect. 4.1 we start with Theorem 4.1 that provides the Gauss-Green formula for any $F \in \mathcal{DM}^1(U)$ and any Borel set $\Omega \subset U$ and also covers specializations for the important cases (G), (L), and (C). Later other special cases are considered. Typical examples show the variability and applicability of the results. In particular, the treatment of singular vector fields, concentrations on the boundaries of open and closed sets, inner boundaries and other degeneracies is demonstrated. The special case of normal measures is considered in Sect. 4.2. The definition and construction of normal measures is followed by some general integrability condition. This leads to Gauss-Green formulas where F and partially also the normal field ν^{Ω} are explicitly contained in the boundary term and where, in addition, some weight on $\partial\Omega$ can be included. Several examples illustrate the variety of applications. Some comprehensive discussion of the results also includes the relation to previous results from the literature. In Sect. 4.3 we briefly transfer the former results for vector fields to Sobolev and BV functions for completeness, but also for the convenience of the reader, since it might be not completely straightforward to do that. In addition we study a Sobolev function on a set Ω of finite perimeter where the trace functional is not related to a Radon measure on $\partial\Omega$ and where both boundary integrals are needed for a general Gauss-Green formula. For a bounded open Ω with Lipschitz boundary, we supplement the classical Gauss-Green formula with a new version that contains a normal measure and does not require a trace function on the boundary. Finally, for any bounded open set Ω the existence of a weak solution for a general boundary value problem is shown.

Summarizing we can say that finitely additive measures appear to be a natural tool for short range phenomena like traces. Though the underlying integration theory was already known for many decades, the bridge to relevant applications seemed to be hidden. Therefore we hope that the new results can somewhat contribute to wake up this "sleeping beauty". The key observation for our investigation was to consider the density $\mathrm{dens}_x A$ as a set function and to realize that the related integral gives the precise representative $\mathcal{L}^n$-a.e. The presented results can certainly be extended to localized spaces $\mathcal{DM}^1_{\mathrm{loc}}(U)$, $\mathcal{BV}_{\mathrm{loc}}(U)$ etc. by using compactly supported test functions φ. But we refrain from formulating the results in that generality in order to avoid unnecessary technicalities for this new approach. Notice that some of the results can already be found in Schönherr [35], and Schönherr-Schuricht [36, 37].

Chapter 2
Preliminaries About Measures

We first provide some material about (finitely additive) measures, as needed for our analysis, that is mostly taken from Bhaskara Rao and Bhaskara Rao [4], Schönherr [35], and Schönherr-Schuricht [36, 37], but also some new preliminary results are contained (notice that our terminology partially differs from that in [4]). Let us also refer to the introduction to finitely additive measures in Dunford and Schwartz [21].

2.1 Measures and Integration

Let Ω be a set and let $\mathcal{A}$ be an *algebra* on Ω, i.e. a collection of subsets of Ω containing $\emptyset$, Ω and being closed under complements and finite unions and intersections. At variance with common usage we call a set function $\mu : \mathcal{A} \to \overline{\mathbb{R}}$ a *measure* on $(\Omega, \mathcal{A})$ if it is finitely additive, i.e.

$$\mu\Big(\bigcup_{k=1}^{m} A_k\Big) = \sum_{k=1}^{m} \mu(A_k)$$

for all pairwise disjoint $A_k \in \mathcal{A}$. We call μ a σ*-measure* on $(\Omega, \mathcal{A})$ if it is σ-additive, i.e.

$$\mu\Big(\bigcup_{k=1}^{\infty} A_k\Big) = \sum_{k=1}^{\infty} \mu(A_k)$$

for any sequence $\{A_k\}$ of pairwise disjoint $A_k \subset \mathcal{A}$ with $\bigcup_{k=1}^{\infty} A_k \in \mathcal{A}$ (usually we denote general measures by μ, λ and σ-measures by σ). Notice that a measure cannot attain both values $\pm\infty$ and that always $\mu(\emptyset) = 0$ if μ is finite somewhere.

F. Schuricht, M. Schönherr, *A Theory of Traces and the Divergence Theorem*,
Lecture Notes in Mathematics 2372, https://doi.org/10.1007/978-3-031-86664-7_2

We say that μ is *positive* if $\mu(A) \geq 0$ for all $A \in \mathcal{A}$ and it is *finite* if $\mu(A) \in \mathbb{R}$ for all $A \in \mathcal{A}$. The *positive* and *negative part* $\mu^{\pm} : \mathcal{A} \to \overline{\mathbb{R}}_{\geq 0}$ of measure μ given by

$$\mu^{\pm}(A) := \sup \left\{ \pm \mu(B) \mid B \subset A,\ B \in \mathcal{A} \right\}$$

and the *total variation* $|\mu| : \mathcal{A} \to \overline{\mathbb{R}}_{\geq 0}$ of μ

$$|\mu| := \mu^{+} + \mu^{-}$$

are positive measures (cf. [4, pp. 53, 85]). A measure μ is *pure* if one has for any σ-measure $\sigma : \mathcal{A} \to \overline{\mathbb{R}}$ that

$$0 \leq \sigma \leq |\mu| \quad \text{implies} \quad \sigma = 0$$

(cf. [4, p. 240]). Thus a (nontrivial) pure measure cannot be extended to a σ-measure on $\mathcal{A}$, but notice that a non σ-additive measure need not be pure. With μ also $\mu^{\pm}$ and $|\mu|$ are pure. The *outer measure* $\mu^* : \mathcal{P}(\Omega) \to [0, \infty]$ of a positive measure μ given by

$$\mu^*(A) := \inf_{\substack{A \subset B \\ B \in \mathcal{A}}} \mu(B)$$

is finitely subadditive but not necessarily σ-subadditive (cf. [4, p. 86]). $A \subset \Omega$ is a *μ-null set* if $|\mu|^*(A) = 0$. A measure μ is *bounded* if

$$\sup_{A \in \mathcal{A}} |\mu(A)| < \infty$$

and it is *bounded above* (*below*) if μ^{+} (μ^{-}) is bounded. We have $\mu = \mu^{+} - \mu^{-}$ if μ^{+} or μ^{-} is bounded. Let us set

$$\begin{aligned}
\mathrm{ba}(\Omega, \mathcal{A}) &:= \{\mu : \mathcal{A} \to \mathbb{R} \mid \mu \text{ is a bounded measure}\}, \\
\mathrm{pa}(\Omega, \mathcal{A}) &:= \{\mu \in \mathrm{ba}(\Omega, \mathcal{A}) \mid \mu \text{ is pure}\}, \\
\mathrm{ca}(\Omega, \mathcal{A}) &:= \{\sigma \in \mathrm{ba}(\Omega, \mathcal{A}) \mid \sigma \text{ is } \sigma\text{-additive}\}
\end{aligned}$$

where $\mathrm{ba}(\Omega, \mathcal{A})$ is a Banach space with $\|\mu\| := |\mu|(\Omega)$ (cf. [4, p. 44]). We call $\mu \in \mathrm{ba}(\Omega, \mathcal{A})^m$ also *vector measure* if $m > 1$. As *total variation* of $\mu \in \mathrm{ba}(\Omega, \mathcal{A})^m$ we define for $A \in \mathcal{A}$

$$|\mu|(A) := \sup \left\{ \sum_{j=1}^{k} |\mu(A_j)| \;\middle|\; A_j \in \mathcal{A} \text{ pairwise disjoint}, \bigcup_{j=1}^{k} A_j = A \right\}. \tag{2.1}$$

Then we have $|\mu| \in \mathrm{ba}(\Omega, \mathcal{A})$ (argue as in the proof of Theorem 1.6 in [2]) and for $m = 1$ this coincides with the previous definition (cf. [4, p. 46]). Obviously $\|\mu\| := |\mu|(\Omega)$ is a norm on $\mathrm{ba}(\Omega, \mathcal{A})^m$ that we use as standard norm. For a Borel set $\Omega \subset \mathbb{R}^n$ a σ-measure on $(\Omega, \mathcal{B}(\Omega))$ is said to be a *Radon measure* if it is finite on compact sets (cf. [22, p. 5]).

Measures μ, λ on $(\Omega, \mathcal{A})$ are called *singular* ($\mu \perp \lambda$) if for all $\varepsilon > 0$ there is some $A \in \mathcal{A}$ with

$$|\mu|(A) < \varepsilon \quad \text{and} \quad |\lambda|(A^c) < \varepsilon$$

and they are called *strongly singular* ($\mu \perp^s \lambda$) if there is some $A \in \mathcal{A}$ with

$$|\mu|(A) = 0 = |\lambda|(A^c)\,.$$

While strong singularity implies singularity, equivalence is met for σ-measures on a σ-algebra (cf. [4, p. 165]). Singularity also means orthogonality, i.e. $|\mu| \wedge |\lambda| = 0$, on the lattice of measures (cf. [4, pp. 52, 166]). For general measures μ we have

$$\mu^+ \perp \mu^- \quad \text{if } \mu^+ \text{ or } \mu^- \text{ is bounded}$$

(cf. [4, p. 53]). Moreover,

$$\mu_c \perp \mu_p \quad \text{if} \quad \mu_c \in \mathrm{ca}(\Omega, \mathcal{A}), \quad \mu_p \in \mathrm{pa}(\Omega, \mathcal{A})$$

(cf. [4, p. 240]). As a consequence of Riesz's decomposition theorem for lattices we have an important singular decomposition of bounded measures (cf. [4, p. 241]).

Proposition 2.1 *Let $\mathcal{A}$ be an algebra on Ω and let $\mu \in \mathrm{ba}(\Omega, \mathcal{A})$. Then there is a unique decomposition into σ*-additive part *μ_c and* pure part *μ_p with*

$$\mu = \mu_c + \mu_p \quad \textit{where} \quad \mu_c \in \mathrm{ca}(\Omega, \mathcal{A})\,, \ \mu_p \in \mathrm{pa}(\Omega, \mathcal{A})\,.$$

By uniqueness we have $\mu^\pm = \mu_c^\pm + \mu_p^\pm$ and, by [4, p. 242], for all $A \in \mathcal{A}$

$$\mu_c^\pm(A) = \inf\left\{ \lim_{k\to\infty} \mu^\pm(A_k) \,\middle|\, A_k \in \mathcal{A}\,,\ A_k \subset A\,,\ A_k \uparrow A \right\}. \tag{2.2}$$

Many examples of pure measures found in the literature are just measures on $\Omega = \mathbb{N}$. Moreover they are either defined on a very small algebra or they rely on some construction not allowing an explicit computation of the measure even on simple sets (cf. [4, p. 246], [51, p. 57 ff.]). Let us provide some simple but typical examples of pure measures.

Example 2.2 For $\Omega = \mathbb{N}$ and the algebra $\mathcal{A} = \{A \subset \Omega \mid A \text{ or } A^c \text{ finite}\}$ we readily get a measure by

$$\mu(A) := \begin{cases} 0 & \text{if } A \text{ is finite,} \\ 1 & \text{if } A^c \text{ is finite} \end{cases}$$

(cf. [4, Remark 4.1.5]). For $\sigma \in \mathrm{ca}(\Omega, \mathcal{A})$ with $0 \le \sigma \le \mu$ we have

$$0 \le \sigma(\mathbb{N}) = \sum_{k \in \mathbb{N}} \sigma(\{k\}) \le \sum_{k \in \mathbb{N}} \mu(\{k\}) = 0$$

and hence μ is pure (cf. also [4, Example 10.4.1]). By $\mu(\mathbb{N}) = 1$ we also see that μ is not σ-subadditive.

Example 2.3 Let $\Omega = (0, 1)$ and let $\mathcal{A}$ be the algebra generated by all intervals of the form $(a, b] \subset \Omega$. Then we get a pure measure by

$$\mu(A) := \begin{cases} 1 & \text{if } B_\delta(0) \cap A \ne \emptyset \text{ for all } \delta > 0\,, \\ 0 & \text{otherwise} \end{cases}$$

(cf. [4, Example 10.4.4]).

Notice that contact interactions in continuum mechanics naturally lead to pure measures having some similarity with that in the last example (cf. Schuricht [38]). In geometric measure theory the density $\mathrm{dens}_x A$ of set A at point x is a well-known important quantity. But it seems that it has not been considered yet as a set function (for fixed x). It turns out to be a typical example of pure measures.

Example 2.4 Let $\Omega := \mathbb{R}^n$, let $\mathcal{A} = \mathcal{B}(\Omega)$, and fix some $x \in \Omega$. The density of $A \in \mathcal{B}(\Omega)$ at x is given by

$$\mathrm{dens}_x A := \lim_{\delta \downarrow 0} \frac{\mathcal{L}^n(A \cap B_\delta(x))}{\mathcal{L}^n(B_\delta(x))}$$

if the limit exists and, as set function, it is disjointly additive on these sets. Though not uniquely, one can extend dens_x to a positive measure on $\mathcal{B}(\Omega)$ such that for all $A \in \mathcal{B}(\Omega)$

$$\liminf_{\delta \downarrow 0} \frac{\mathcal{L}^n(A \cap B_\delta(x))}{\mathcal{L}^n(B_\delta(x))} \le \mathrm{dens}_x A \le \limsup_{\delta \downarrow 0} \frac{\mathcal{L}^n(A \cap B_\delta(x))}{\mathcal{L}^n(B_\delta(x))} \tag{2.3}$$

(cf. Proposition 3.1 below with $U = E = \Omega$, $\Gamma = \{x\}$, $\gamma(\delta) = \mathcal{L}^n(B_\delta(x))$, $f \equiv 1$, and $\varphi = \chi_A$). We call dens_x *density measure* at x. For $\sigma \in \mathrm{ca}(\Omega, \mathcal{A})$ with $0 \le \sigma \le \mathrm{dens}_x$ we get

$$0 \le \sigma\big(\mathbb{R}^n \setminus \{x\}\big) = \lim_{\delta \downarrow 0} \sigma\big(B_\delta(x)^c\big) \le \lim_{\delta \downarrow 0} \mathrm{dens}_x\big(B_\delta(x)^c\big) = 0$$

and $0 \le \sigma(\{x\}) \le \text{dens}_x(\{x\}) = 0$. Hence $\sigma = 0$ and dens_x is pure. Moreover $\text{dens}_x \perp \mathcal{L}^n$ but not $\text{dens}_x \perp^s \mathcal{L}^n$.

For measures λ, μ on algebra $\mathcal{A}$, we call μ *absolutely continuous* with respect to λ and write $\mu << \lambda$, if for every $\varepsilon > 0$ there is some $\delta > 0$ such that

$$|\lambda|(A) < \delta \quad \text{implies} \quad |\mu(A)| < \varepsilon \quad \text{for} \quad A \in \mathcal{A}\,.$$

Then also $|\mu| << \lambda$. We call μ *weakly absolutely continuous* with respect to λ if

$$|\lambda|(A) = 0 \quad \text{implies} \quad \mu(A) = 0 \quad \text{for} \quad A \in \mathcal{A} \tag{2.4}$$

and we write $\mu <<^w \lambda$. Then also $|\mu| <<^w \lambda$, since (2.4) is valid for all $A' \subset A$ if it is true for A. Clearly $\mu <<^w \lambda$ if $\mu << \lambda$ and both notions coincide for positive bounded σ-measures on a σ-algebra (cf. [4, p. 159 ff.]). For any measure λ on $(\Omega, \mathcal{A})$ (λ not necessarily bounded) we introduce the set

$$\text{ba}(\Omega, \mathcal{A}, \lambda) := \big\{\mu \in \text{ba}(\Omega, \mathcal{A}) \mid \mu <<^w \lambda\big\}$$

(the space $\text{ba}(\Omega, \mathcal{B}(\Omega), \mathcal{L}^n)$ will play an important role in our subsequent analysis). For $\mu \in \text{ba}(\Omega, \mathcal{A}, \lambda)$ with $\mu = \mu_c + \mu_p$ according to Proposition 2.1, we get $\mu_c \in \text{ba}(\Omega, \mathcal{A}, \lambda)$ if we apply (2.2) to A with $\lambda(A) = 0$. Then $\mu_p \in \text{ba}(\Omega, \mathcal{A}, \lambda)$, since it is a linear space (cf. also [37, Theorem 3.16]).

Proposition 2.5 *Let $\mathcal{A}$ be an algebra on Ω, let λ be a measure on $(\Omega, \mathcal{A})$, and assume that $\mu \in \text{ba}(\Omega, \mathcal{A}, \lambda)$ with (unique) decomposition $\mu = \mu_c + \mu_p$ according to Proposition* 2.1. *Then*

$$\mu_c, \mu_p \in \text{ba}(\Omega, \mathcal{A}, \lambda).$$

The following characterization of pure measures in $\text{ba}(\Omega, \mathcal{A}^\sigma, \sigma)$ with σ being a σ-measure adumbrates their relation to traces (cf. [4, p. 244] and [51, p. 56]).

Proposition 2.6 *Let $\mathcal{A}^\sigma$ be a σ-algebra on Ω and let $\sigma \in \text{ca}(\Omega, \mathcal{A}^\sigma)$ be positive. Then $\mu \in \text{ba}(\Omega, \mathcal{A}^\sigma, \sigma)$ is pure if and only if there is a decreasing sequence $\{A_k\} \subset \mathcal{A}^\sigma$ such that*

$$\sigma(A_k) \to 0 \quad \textit{and} \quad |\mu|(A_k^c) = 0 \ \textit{for all} \ k \in \mathbb{N}\,.$$

Roughly speaking one "feels" a pure measure $\mu \in \text{ba}(\Omega, \mathcal{A}^\sigma, \sigma)$ in any small vicinity of a σ-null set $\bigcap_{k \in \mathbb{N}} A_k$. We call $A \in \mathcal{A}$ an *aura* of $\mu \in \text{ba}(\Omega, \mathcal{A})$ if

$$|\mu|(A^c) = 0\,.$$

For a pure measure $\mu \in \text{ba}(\Omega, \mathcal{A}^\sigma, \sigma)$ a decreasing sequence $\{A_k\}$ of auras with $\sigma(A_k) \to 0$ is said to be an *aura sequence* for μ. Notice that the intersection of an aura sequence might be empty. Hence the support as used for σ-measures on Ω with a topological structure (that identifies a preferably small set where the complete mass of the measure concentrates, cf. [2, p. 30]) might be empty for a nontrivial measure. In fact, for the pure measure dens_x from Example 2.4 the support would be $\{x\}$, but

$$\text{dens}_x(\{x\}) = 0 \quad \text{and} \quad \text{dens}_x(B_\delta(x) \setminus \{x\}) = 1 \text{ for all } \delta > 0 .$$

Notice that we can define dens_x merely on $\Omega = \mathbb{R}^n \setminus \{x\}$ which leads to a support outside of Ω. Thus we have to realize that the usual support cannot localize where a general measure "lives". For a subset $\Omega \subset M$ of a compact topological space M and an algebra $\mathcal{A}$ containing all relatively open sets in Ω, we introduce the *core* of $\mu \in \text{ba}(\Omega, \mathcal{A})$ as the set

$$\text{core}\, \mu := \big\{x \in M \;\big|\; |\mu|(U \cap \Omega) > 0 \text{ for all } U \subset M \text{ open with } x \in U\big\} .$$

Obviously $\text{core}\, \mu \subset \overline{\Omega}$, it is closed in M, and it need not be contained in Ω. For $\mu \neq 0$ one has $\text{core}\, \mu \neq \emptyset$ and

$$|\mu|(U^c \cap \Omega) = 0 \quad \text{for any open} \quad U \subset M \quad \text{with} \quad \text{core}\, \mu \subset U \tag{2.5}$$

(cf. Proposition 2.11 below). From $\text{core}(\text{dens}_x) = \{x\}$ we see that a nonempty core might be a μ-null set. For $\Omega = \mathbb{R}^n$ or $\Omega = \mathbb{N}$ the core is not defined, since Ω is not compact (with the usual topology). Notice that formally $\text{core}\, \mu = \emptyset$ for the non-zero measure μ from Example 2.2 and we would sloppily say that μ "concentrates near ∞". We can describe such situations precisely by (tacitly) using the compactifications

$$\widehat{\mathbb{N}} := \mathbb{N} \cup \{\infty\} \quad \text{and} \quad \widehat{\mathbb{R}^n} := \mathbb{R}^n \cup \{\infty\}$$

in the definition of core. This way we get $\text{core}\, \mu = \{\infty\}$ for μ from Example 2.2 and, e.g., for any $\mu \neq 0$ on $\mathbb{R}^n$ with $\mu(A) = 0$ for all bounded $A \subset \mathbb{R}^n$ (cf. also [4, Example 10.4.1]). For $\Omega \in \mathcal{B}(\mathbb{R}^n)$ and $\mu \in \text{ba}(\Omega, \mathcal{B}(\Omega), \mathcal{L}^n)$ we have that

$$\mu \quad \text{is pure if} \quad \mathcal{L}^n(\text{core}\, \mu \cap \Omega) = 0 .$$

(cf. Proposition 2.12 below). It turns out that core and aura are reasonable tools to describe where the measure is concentrated.

For a measure μ on $(\Omega, \mathcal{A})$, a sequence of functions $f_k : \Omega \to \mathbb{R}$ *converges in measure* to $f : \Omega \to \mathbb{R}$ if

$$\lim_{k\to\infty} |\mu|^*\big(\big\{x \in \Omega \;\big|\; |f_k(x) - f(x)| > \varepsilon\big\}\big) = 0 \quad \text{for all} \quad \varepsilon > 0$$

and we write $f_k \xrightarrow{\mu} f$. The limit is unique if we identify functions $f, g : \Omega \to \mathbb{R}$ that *agree in measure* μ (i.m.) on Ω in the sense that

$$|\mu|^*\big(\big\{x \in \Omega \;\big|\; |f(x) - g(x)| > \varepsilon\big\}\big) = 0 \quad \text{for every} \quad \varepsilon > 0$$

(cf. [4, pp. 88, 92]). The stronger condition that f and g *agree* μ *-almost everywhere* (a.e.) on Ω, i.e.

$$|\mu|^*\big(\big\{x \in \Omega \;\big|\; |f(x) - g(x)| \neq 0\big\}\big) = 0\,,$$

is sufficient but not necessary for $f = g$ i.m. in general, however it is equivalent in the case of a σ-measure μ on a σ-algebra $\mathcal{A}$ (cf. [4, Proposition 4.2.7]). For demonstration we consider the measure $\mu = \mathrm{dens}_0$ from Example 2.4 and the functions $f_c(x) = c|x|$ on $\mathbb{R}^n$ with $c \in \mathbb{R}$. Then $f_c = f_d$ i.m. on $\mathbb{R}^n$ for all $c, d \in \mathbb{R}$. But f_c and f_d do not agree a.e. on $\mathbb{R}^n$ for $c \neq d$, since

$$|\mu|^*\big(\big\{x \;\big|\; |f_c(x) - f_d(x)| \neq 0\big\}\big) = \mu\big(\mathbb{R}^n \setminus \{0\}\big) = 1\,.$$

For $f, g : \Omega \to \mathbb{R}$ we say $f \le g$ *in measure* (i.m.) on Ω if

$$|\mu|^*\big(\big\{x \in \Omega \;\big|\; g(x) - f(x) < -\varepsilon\big\}\big) = 0 \quad \text{for every} \quad \varepsilon > 0$$

(or, equivalently, $f \le g + h$ on Ω for some h with $h = 0$ i.m. on Ω, cf. [4, p. 88]). The condition $f \le g$ *almost everywhere* (a.e.) on Ω, i.e.

$$|\mu|^*\big(\big\{x \in \Omega \;\big|\; g(x) - f(x) < 0\big\}\big) = 0\,,$$

is sufficient but not necessary for $f \le g$ i.m. in general. Clearly $f = g$ i.m. if $f \le g$ and $g \le f$ i.m. on Ω (cf. [4, p. 88]).

We call $h : \Omega \to \mathbb{R}$ *simple function* related to measure μ on $(\Omega, \mathcal{A})$ if there are finitely many $c_k \in \mathbb{R}$ and $A_k \in \mathcal{A}$ such that

$$h = \sum_{k=1}^{m} c_k \chi_{A_k} \quad \text{on} \quad \Omega\,.$$

The simple function h is *integrable* (with respect to μ) if $|\mu|(A_k) < \infty$ whenever $c_k \neq 0$ and we set

$$\int_\Omega h\, d\mu := \sum_{k=1}^{m} c_k \mu(A_k)$$

(with the convention $0 \cdot \pm\infty = 0$, cf. [4, p. 96 ff.]).

We say that $f : \Omega \to \mathbb{R}$ is *measurable* (with respect to μ) if there is a sequence of simple functions $h_k : \Omega \to \mathbb{R}$ such that

$$h_k \overset{\mu}{\to} f\,.$$

Then f is measurable if and only if for any $\varepsilon > 0$ there is a partition $A_0, \dots, A_m$ of Ω in $\mathcal{A}$ such that

$$|\mu|(A_0) < \varepsilon \quad \text{and} \quad |f(x) - f(x')| < \varepsilon \ \text{ for all } \ x, x' \in A_k\,,\ k = 1, \dots, m$$

(cf. [4, p. 101]).

We call $f : \Omega \to \mathbb{R}$ *integrable* (or μ-integrable) on Ω if there is a sequence of integrable simple functions $h_k : \Omega \to \mathbb{R}$ such that

$$h_k \overset{\mu}{\to} f \quad \text{and} \quad \lim_{k,l\to\infty} \int_\Omega |h_k - h_l|\, d|\mu| = 0\,.$$

In this case f is measurable and we define the *integral* of f on Ω as

$$\int_\Omega f\, d\mu := \lim_{k\to\infty} \int_\Omega h_k\, d\mu$$

while $\{h_k\}$ is called a *determining sequence* for it (cf. [4, p. 104]). The integral is linear and integrability of f is equivalent to that of $|f|$ (cf. [4, p. 113]). An integrable f is also integrable with respect to $|\mu|$ and it is integrable on any $A \in \mathcal{A}$. Integrable functions $f, g : \Omega \to \mathbb{R}$ agree i.m. on Ω if and only if

$$\int_A f\, d\mu = \int_A g\, d\mu \quad \text{for all} \quad A \in \mathcal{A}\,.$$

If f is integrable, then

$$\lambda(A) = \int_A f\, d\mu \quad \text{for} \quad A \in \mathcal{A} \tag{2.6}$$

gives a measure $\lambda \in \mathrm{ba}(\Omega, \mathcal{A})$ that is absolutely continuous with respect to μ. For more basic properties we refer to [4, pp. 105–107]. If $f_k, g : \Omega \to \mathbb{R}$ are integrable with

$$f_k \overset{\mu}{\to}: f \quad \text{and} \quad |f_k| \le g \quad \text{i.m. on } \Omega\,,$$

we have dominated convergence, i.e.

$$\lim_{k\to\infty} \int_\Omega f_k\, d\mu = \int_\Omega f\, d\mu$$

(cf. [4, p. 131]). For a vector measure $\mu = (\mu_1, \ldots, \mu_m)$ a scalar function $f : \Omega \to \mathbb{R}$ is said to be μ-integrable if f is μ_j-integrable for all j. A vector-valued function $F = (F_1, \ldots, F_m) : \Omega \to \mathbb{R}^m$ is said to be μ integrable if each F_j is μ_j-integrable and we set

$$\int_\Omega F\, d\mu := \sum_{j=1}^{m} \int_\Omega F_j\, d\mu_j\,.$$

For the integral $\int_\Omega f\, d\mu$ it is sufficient to integrate on an aura $A \subset \Omega$ of μ, but it is not enough to integrate on core μ (if it is defined). For dens_x from Example 2.4, e.g., we have for f continuous and $\delta > 0$ that

$$\int_{B_\delta(x)} f\, d\,\mathrm{dens}_x = f(x) \quad \text{but} \quad \int_{\{x\}} f\, d\,\mathrm{dens}_x = 0$$

(where $B_\delta(x)$ is an aura and $\{x\}$ the core of μ). In order to indicate a more precise information about the domain of integration, we use the notation (if $\operatorname{core}\mu$ is defined)

$$\fint_C f\, d\mu := \int_{U\cap\Omega} f\, d\mu \quad \text{if} \quad \operatorname{core}\mu \subset C \subset U\,, \quad U \text{ open}\,,$$

which is well-defined by (2.5).

Notice that the usual notion of measurable functions, that is based on a σ-measure σ on $(\Omega, \mathcal{A}^\sigma)$ with σ-algebra $\mathcal{A}^\sigma$, relies on convergence a.e. and it is weaker than the one used here in general, but both agree if $\sigma(\Omega) < \infty$. Nevertheless integrability and the integral as introduced here always agree with the usual σ-variant, since usual convergence in $\mathcal{L}^1(\Omega)$ implies convergence in measure due to Chebyshev's inequality.

For a measure μ on $(\Omega, \mathcal{A})$ and $1 \le p < \infty$ we define

$$L^p(\Omega, \mathcal{A}, \mu) := \big\{ f : \Omega \to \mathbb{R} \;\big|\; f \text{ measurable},\ |f|^p \text{ integrable} \big\}$$

with

$$\|f\|_p := \Big(\int_\Omega |f|^p\, d|\mu| \Big)^{\frac{1}{p}}$$

and

$$L^\infty(\Omega, \mathcal{A}, \mu) := \big\{ f : \Omega \to \mathbb{R} \;\big|\; f \text{ measurable},\ \|f\|_\infty < \infty \big\}$$

where

$$\|f\|_\infty := \operatorname*{ess\,sup}_{\Omega} |f| := \inf_{\substack{N \subset \Omega \\ |\mu|^*(N)=0}} \ \sup_{\Omega \setminus N} |f| \,.$$

For vector-valued functions $F \in L^p(\Omega, \mathcal{A}, \mu)^m$ we replace $|f|$ with the Euclidean norm $|F|$ in the previous definitions. The corresponding sets of equivalence classes with respect to equality i.m., denoted by

$$\mathcal{L}^p(\Omega, \mathcal{A}, \mu)\,,$$

are normed spaces but they are not complete in general (cf. [4, p. 125 ff.]).

With $f \in \mathcal{L}^p(\Omega, \mathcal{A}, \mu)$ we get a measure $f\mu \in \mathrm{ba}(\Omega, \mathcal{A})$, $f\mu << \mu$ by

$$(f\mu)(A) = \int_A f \, d\mu \quad \text{for all} \quad A \in \mathcal{A}\,. \tag{2.7}$$

For positive $\mu \in \mathrm{ba}(\Omega, \mathcal{A})$ the completion of $\mathcal{L}^p(\Omega, \mathcal{A}, \mu)$ for $p \in [1, \infty]$ is

$$\mathscr{L}^p(\Omega, \mathcal{A}, \mu) := \big\{\lambda \in \mathrm{ba}(\Omega, \mathcal{A}) \;\big|\; \lambda << \mu\,,\; \|\lambda\|_p < \infty\big\}$$

with

$$\|\lambda\|_p^p := \lim_{\mathcal{P} \in \mathscr{P}} \sum_{\substack{A \in \mathcal{P} \\ \mu(A) \neq 0}} \left|\frac{\lambda(A)}{\mu(A)}\right|^p \mu(A) = \int_\Omega^{\mathcal{R}} \left|\frac{\lambda}{\mu}\right|^p \mu \qquad \big(1 \leq p < \infty\big)\,,$$

where $\int^{\mathcal{R}}$ is the refinement integral (cf. [28], [4, p. 231]) and the limit is taken in the sense of nets over the directed set $\mathscr{P}$ of all finite partitions $\mathcal{P} \subset \mathcal{A}$ of Ω, and

$$\|\lambda\|_\infty := \sup\left\{\left|\frac{\lambda(A)}{\mu(A)}\right| \;\middle|\; A \in \mathcal{A}\right\}$$

(use convention $\frac{0}{0} = 0$, cf. [4, p. 185]). Hölder's inequality is satisfied in all spaces for $1 \leq p \leq \infty$ (cf. [4, p. 122]) and the integrable simple functions are dense in all spaces for $1 \leq p < \infty$ (cf. [4, p. 132]). We briefly write

$$L^p(\Omega, \mu) := L^p(\Omega, \mathcal{B}(\Omega), \mu)\,, \quad L^p(\Omega) := L^p(\Omega, \mathcal{B}(\Omega), \mathcal{L}^n)\,,$$

for vector-valued functions we write

$$L^p(\Omega, \mathcal{B}(\Omega), \mu)^m\,, \quad L^p(\Omega, \mu)^m\,, \quad L^p(\Omega, \mathbb{R}^m) := L^p(\Omega)^m\,,$$

and we use an analogous notation for $\mathcal{L}^p$-spaces. The dual of $\mathcal{L}^\infty(\Omega, \mathcal{A}^\sigma, \sigma)$ can be identified with $\mathrm{ba}(\Omega, \mathcal{A}^\sigma, \sigma)$ and plays an important role in our analysis (cf.

[4, pp. 139, 140] or also [51, p. 53], [50, p. 118]). Let us formulate a vector-valued version as needed later. The more refined decomposition, that makes precise how $\mathcal{L}^1(\Omega, \mathcal{A}^\sigma, \sigma)$ as subspace of the dual has to be supplemented, relies on Proposition 2.5 and the Radon-Nikodym theorem.

Proposition 2.7 *Let σ be a positive σ-measure on $(\Omega, \mathcal{A}^\sigma)$ with σ-algebra $\mathcal{A}^\sigma$. Then, for $m \in \mathbb{N}$,*

$$\left(\mathcal{L}^\infty(\Omega, \mathcal{A}^\sigma, \sigma)^m\right)^* = \mathrm{ba}(\Omega, \mathcal{A}^\sigma, \sigma)^m$$

if we identify $f^* \in \left(\mathcal{L}^\infty(\Omega, \mathcal{A}^\sigma, \sigma)^m\right)^*$ and $\mu \in \mathrm{ba}(\Omega, \mathcal{A}^\sigma, \sigma)^m$ by

$$\langle f^*, \varphi\rangle = \int_\Omega \varphi \, d\mu \qquad \text{for all} \quad \varphi \in \mathcal{L}^\infty(\Omega, \mathcal{A}^\sigma, \sigma)^m$$

where

$$\|f^*\| = \|\mu\| = |\mu|(\Omega) \tag{2.8}$$

and

$$|\mu|(A) = \sup_{\substack{\varphi \in \mathcal{L}^\infty(\Omega, \mathcal{A}^\sigma, \sigma)^m \\ \|\varphi\|_\infty \le 1}} \int_A \varphi \, d\mu \quad \text{for all} \ A \in \mathcal{A}^\sigma . \tag{2.9}$$

Moreover, a measure $\mu \in \mathrm{ba}(\Omega, \mathcal{A}^\sigma, \sigma)^m$ can be decomposed uniquely with some pure $\mu_p \in \mathrm{ba}(\Omega, \mathcal{A}^\sigma, \sigma)^m$ and some $h \in \mathcal{L}^1(\Omega, \mathcal{A}^\sigma, \sigma)^m$ such that

$$\int_\Omega f \, d\mu = \int_\Omega f \cdot h \, d\sigma + \int_\Omega f \, d\mu_p \qquad \text{for all} \quad f \in \mathcal{L}^\infty(\Omega, \mathcal{A}^\sigma, \sigma)^m ,$$

i.e. $\mu = \mu_c + \mu_p$ with $\mu_c = h\sigma$ according to Proposition 2.5.

Notice that (2.9) readily follows from (2.1) for $m = 1$. Let us here also refer to the very comprehensive investigation of the dual $(\mathcal{L}^\infty)^*$ with focus on weak convergence in $\mathcal{L}^\infty$ by Toland [46].

Proof For all assertions except (2.9) it is sufficient to consider the case $m = 1$. As already mentioned, the characterization of the dual space in the scalar case follows from [4, pp. 139, 140]. The decomposition relies on Proposition 2.5 and the Radon-Nikodym theorem (cf. also [37, Theorem 4.14]).

It remains to show (2.9) for a vector measure $\mu = (\mu_1, \ldots, \mu_m)$ (cf. [2, p. 21] for the case of a σ-measure). For that we fix $\varepsilon > 0$. First, by (2.1), there is a pairwise disjoint decomposition $A = \bigcup_{j=1}^{k} A_j$ such that $\mu(A_j) \neq 0$, $A_j \in A^\sigma$, and

$$|\mu|(A) - \varepsilon \leq \sum_{j=1}^{k} |\mu(A_j)| = \sum_{j=1}^{k} \int_{A_j} a_j \, d\mu = \int_A \tilde{\varphi} \, d\mu \leq \sup_{\substack{\varphi \in \mathcal{L}^\infty(\Omega, \mathcal{A}^\sigma, \sigma)^m \\ \|\varphi\|_{\mathcal{L}^\infty} \leq 1}} \int_A \varphi \, d\mu$$

where $a_j = \frac{\mu(A_j)}{|\mu(A_j)|}$ and $\tilde{\varphi} = a_j$ on A_j. Second, there is some $\tilde{\varphi} \in \mathcal{L}^\infty(\Omega, \mathcal{A}^\sigma, \sigma)^m$ with $\|\tilde{\varphi}\|_\infty \leq 1$ and there is a step function $h = \sum_{j=1}^{k} a_j \chi_{A_j}$ ($A_j \in A^\sigma$ pairwise disjoint) with $|\tilde{\varphi} - h| < \varepsilon$, $|h| \leq 1$ on A such that

$$\begin{aligned} \sup_{\substack{\varphi \in \mathcal{L}^\infty(\Omega, \mathcal{A}^\sigma, \sigma)^m \\ \|\varphi\|_{\mathcal{L}^\infty} \leq 1}} \int_A \varphi \, d\mu - \varepsilon \leq \int_A \tilde{\varphi} \, d\mu \leq \int_A h \, d\mu + c\varepsilon \\ = \sum_{j=1}^{k} \int_{A_j} a_j \, d\mu + c\varepsilon \leq \sum_{j=1}^{k} |\mu(A_j)| + c\varepsilon \\ \leq |\mu|(A) + c\varepsilon \end{aligned}$$

where $c = |\mu|(A)$. Since $\varepsilon > 0$ is arbitrary, the assertion follows. □

Now we provide some examples for integration.

Example 2.8 With μ on $(\mathbb{N}, \mathcal{A})$ from Example 2.2 and functions on $\mathbb{N}$ taken as sequences $\{a_n\} \subset \mathbb{R}$, we have integrability if and only if there is some $c \in \mathbb{R}$ with $a_k = c$ for almost all $k \in \mathbb{N}$ and then

$$\int_{\mathbb{N}} a_k \, d\mu = c$$

(cf. [4, p. 111]).

Example 2.9 The space $\ell^\infty := \mathcal{L}^\infty(\mathbb{N}, \mathcal{P}(\mathbb{N}), \sigma)$ with the counting measure σ is just the set of bounded sequences $\{a_k\} \subset \mathbb{R}$. On the subspace ℓ_0^∞ of convergent sequences one has a linear continuous functional by

$$\{a_k\} \to \lim_k a_k$$

which can be extended (not uniquely) on ℓ^∞ by Hahn-Banach. Hence there is some $\mu \in \mathrm{ba}(\mathbb{N}, \mathcal{P}(\mathbb{N}), \sigma)$ such that all sequences of ℓ^∞ are integrable with respect to μ and

$$\int_{\mathbb{N}} a_k \, d\mu = \lim_k a_k \quad \text{on} \quad \ell_0^\infty$$

(cf. also [4, p. 39 ff.] and Banach limits in [50, p. 104]). Sequences $\{a_k\}$ that are zero except for some $a_l = 1$ are simple functions on $\mathbb{N}$ and we get

$$\mu(\{l\}) = \int_{\mathbb{N}} a_k \, d\mu = 0 \quad \text{for all} \quad l \in \mathbb{N}\,.$$

Hence μ is pure by arguments as in Example 2.2.

Example 2.10 Obviously dens_x from Example 2.4 belongs to $\mathrm{ba}(\mathbb{R}^n, \mathcal{B}(\mathbb{R}^n), \mathcal{L}^n)$. Hence all functions $f \in L^\infty(\mathbb{R}^n)$ are integrable with respect to dens_x for any $x \in \mathbb{R}^n$. Proposition 3.1 below implies that

$$\bar{f}(x) := \int_{\mathbb{R}^n} f \, d\, \mathrm{dens}_x = \fint_{\{x\}} f \, d\, \mathrm{dens}_x$$

agrees with $f(x)$ at Lebesgue points x of f, since there

$$f(x) = \lim_{\delta \downarrow 0} \fint_{B_\delta(x)} f \, d\mathcal{L}^n\,,$$

while $\bar{f}(x)$ may depend on the special choice of the measure dens_x at other points. This way we have an integral representation for some $\bar{f} \in L^\infty(\mathbb{R}^n)$ at all $x \in \mathbb{R}^n$ that agrees $\mathcal{L}^n$-a.e. with the precise representative of f (cf. also Remark 2.16 below).

Let us justify the definition of core before we consider some facts about weakly absolutely continuous measures.

Proposition 2.11 *Let M be a compact topological space, let $\Omega \subset M$, let $\mathcal{A}$ be an algebra containing all relatively open sets in Ω, and let $\mu \in \mathrm{ba}(\Omega, \mathcal{A})$ with $\mu \neq 0$. Then* core $\mu \neq \emptyset$ *and*

$$|\mu|(U \cap \Omega) = |\mu|(\Omega)\,, \quad |\mu|(U^c \cap \Omega) = 0$$

for any open $U \subset M$ with core $\mu \subset U$.

Proof Assume that core $\mu = \emptyset$. Then, by definition of core and by compactness of M, there is a finite open covering $\{U_k\}_{k=1}^m$ of M with $|\mu|(U_k \cap \Omega) = 0$ for all k. Hence,

$$|\mu|(\Omega) \leq \sum_{k=1}^{m} |\mu|\big(U_k \cap \Omega\big) = 0$$

in contradiction to $\mu \neq 0$.

Now let $U \subset M$ be open such that core $\mu \subset U$. Then, for any $y \in U^c$, we find some open neighborhood U_y with $|\mu|(U_y \cap \Omega) = 0$ by definition of core. Since U^c is compact, the open covering $\{U_y\}_{y \in U^c}$ contains a finite covering $\{U_j\}_{j=1}^l$.

Consequently,

$$0 \leq |\mu|(U^c \cap \Omega) \leq \sum_{j=1}^{l} |\mu|(U_j \cap \Omega) = 0$$

and

$$|\mu|(\Omega) = |\mu|(U \cap \Omega) + |\mu|(U^c \cap \Omega) = |\mu|(U \cap \Omega)$$

which finishes the proof. □

2.2 Some Weakly Absolutely Continuous Measures

The measures in $\mathrm{ba}(\Omega, \mathcal{B}(\Omega), \mathcal{L}^n)$, that are weakly absolutely continuous with respect to $\mathcal{L}^n$ and that coincide with the linear continuous functionals on $\mathcal{L}^\infty(\Omega, \mathcal{B}(\Omega), \mathcal{L}^n)$, are of particular interest for our general treatment of traces. Therefore let us discuss some special aspects. First we provide a very useful sufficient condition for such measures to be pure.

Proposition 2.12 *Let $\Omega \in \mathcal{B}(\mathbb{R}^n)$. Then $\mu \in \mathrm{ba}(\Omega, \mathcal{B}(\Omega), \mathcal{L}^n)$ is pure if*

$$\mathcal{L}^n(\mathrm{core}\,\mu \cap \Omega) = 0\,.$$

Proof We have that $\mathrm{core}\,\mu$ is closed in the compact topological space $\widehat{\mathbb{R}^n}$ and there is a decreasing sequence B_k of open neighborhoods of $\mathrm{core}\,\mu$ such that $\mathrm{core}\,\mu = \bigcap_k B_k$. Then

$$|\mu|\big(\Omega \setminus B_k\big) = 0 \quad \text{for all } \; k \in \mathbb{N}$$

by Proposition 2.11. For a positive σ-measure σ with $0 \leq \sigma \leq |\mu|$ we get

$$0 \leq \sigma(\Omega \setminus B_k) \leq |\mu|(\Omega \setminus B_k) = 0\,.$$

From $\mathcal{L}^n(\mathrm{core}\,\mu \cap \Omega) = 0$ we derive

$$0 \leq \sigma(\mathrm{core}\,\mu \cap \Omega) \leq |\mu|(\mathrm{core}\,\mu \cap \Omega) = 0\,.$$

Therefore

$$\sigma(\Omega) = \sigma(\mathrm{core}\,\mu \cap \Omega) + \sigma\Big(\bigcup_k \Omega \setminus B_k\Big) = \lim_{k \to \infty} \sigma\big(\Omega \setminus B_k\big) = 0$$

and thus, $\sigma = 0$. Since σ was arbitrary, μ is pure. □

Now we show how any measure $\mu \in \mathrm{ba}(\Omega, \mathcal{B}(\Omega))$ can be "restricted" to a Radon measure supported on core μ and how a σ-measure on some $\Gamma \subset \overline{\Omega}$ can be extended to a measure $\mu \in \mathrm{ba}(\Omega, \mathcal{B}(\Omega), \mathcal{L}^n)$ with core $\mu \subset \overline{\Gamma}$. For the proof of the second statement we use the semi norm on $\mathcal{L}^\infty(\Omega, \mathbb{R}^m)$ given by

$$\|\varphi\|_\Gamma := \lim_{\delta \downarrow 0} \|\varphi\|_{\mathcal{L}^\infty(\Gamma_\delta \cap \Omega)} = \inf_{\delta > 0} \|\varphi\|_{\mathcal{L}^\infty(\Gamma_\delta \cap \Omega)} \tag{2.10}$$

which is well-defined for Borel sets Γ and Ω with $\Gamma \subset \overline{\Omega}$ (the limit exists and equals the infimum, since the norm on the right hand side is increasing in δ).

Proposition 2.13 *Let $\Omega \subset \mathbb{R}^n$ be a bounded Borel set.*

(1) *For $\mu \in \mathrm{ba}(\Omega, \mathcal{B}(\Omega))^m$ there is a Radon measure $\sigma \in \mathrm{ca}(\overline{\Omega}, \mathcal{B}(\overline{\Omega}))^m$ supported on* core μ *such that*

$$\int_{\mathrm{core}\,\mu} \varphi \, d\sigma = \int_\Omega \varphi \, d\mu \quad \Big(= \fint_{\mathrm{core}\,\mu} \varphi \, d\mu \Big) \qquad \textit{for all} \quad \varphi \in C(\overline{\Omega}, \mathbb{R}^m)\,.$$

(2) *Let $\sigma \in \mathrm{ca}(\Gamma, \mathcal{B}(\Gamma))^m$ be a bounded vector-valued σ-measure on some Borel set $\Gamma \subset \overline{\Omega}$ such that*

$$\mathcal{L}^n(B_\delta(x) \cap \Omega) > 0 \qquad \textit{for all} \quad x \in \Gamma,\ \delta > 0\,. \tag{2.11}$$

Then there exists $\mu \in \mathrm{ba}(\Omega, \mathcal{B}(\Omega), \mathcal{L}^n)^m$ with

$$\mathrm{core}\,\mu \subset \overline{\Gamma} \quad \textit{and} \quad |\mu|(\Omega) = |\sigma|(\Gamma)$$

such that

$$\int_\Gamma \varphi \, d\sigma = \int_\Omega \varphi \, d\mu \quad \Big(= \fint_{\overline{\Gamma}} \varphi \, d\mu \Big) \qquad \textit{for all} \quad \varphi \in C(\overline{\Omega}, \mathbb{R}^m)\,.$$

If $\mathcal{L}^n(\overline{\Gamma} \cap \Omega) = 0$, then μ is pure.

Obviously (2.11) is satisfied if e.g. $\Gamma \subset \partial_* \Omega \cup \mathrm{int}_* \Omega$. The construction of measure μ in (2) uses the Hahn-Banach theorem and is not unique. Before giving the proof we consider some example as illustration.

Example 2.14 Let $\Omega \subset \mathbb{R}^n$ be open and bounded, let $\Gamma = \{x\}$ for some $x \in \Omega$, and let δ_x be the Dirac measure concentrated at x. By Proposition 2.13 there is some pure $\mu_x \in \mathrm{ba}(\Omega, \mathcal{B}(\Omega), \mathcal{L}^n)$ such that

$$\varphi(x) = \int_{\{x\}} \varphi \, d\delta_x = \fint_{\{x\}} \varphi \, d\mu_x \qquad \text{for all} \quad \varphi \in C(\Omega)\,. \tag{2.12}$$

Obviously $\mu_x = \text{dens}_x$ would be a possible choice (cf. Example 2.10). Alternatively we can consider some density measure $\text{dens}_x^E \in \text{ba}(\Omega, \mathcal{B}(\Omega), \mathcal{L}^n)$ at x with respect to $E \in \mathcal{B}(\Omega)$ by a construction in analogy to Example 2.4 such that

$$\liminf_{\delta \downarrow 0} \frac{\mathcal{L}^n(A \cap B_\delta^E(x))}{\mathcal{L}^n(B_\delta^E(x))} \le \text{dens}_x^E(A) \le \limsup_{\delta \downarrow 0} \frac{\mathcal{L}^n(A \cap B_\delta^E(x)}{\mathcal{L}^n(B_\delta^E(x))} \tag{2.13}$$

where $B_\delta^E(x) = B_\delta(x) \cap E$ (Proposition 3.1 below ensures the existence of such measures). We call dens_x^E a *density measure* at x with respect to E. In particular we can choose $\mu_x = \text{dens}_x^E$ for some open E having an outward cusp at $x \in \partial E$. Then (2.12) remains true with $\mu_x = \text{dens}_x^E$ (cf. Proposition 3.1). But, for $\varphi = \chi_E \in L^\infty(\Omega)$ we have that x is Lebesgue point and

$$\begin{aligned} 0 = \chi_E(x) &= \lim_{\delta \downarrow 0} \fint_{B_\delta(x)} \chi_E \, d\mathcal{L}^n = \text{dens}_x(E) = \fint_{\{x\}} \chi_E \, d\,\text{dens}_x \\ \ne 1 &= \lim_{\delta \downarrow 0} \fint_{E \cap B_\delta(x)} \chi_E \, d\mathcal{L}^n = \text{dens}_x^E(E) = \fint_{\{x\}} \chi_E \, d\,\text{dens}_x^E . \end{aligned}$$

Hence a point evaluation $\bar{\varphi}$ with dens_x^E in analogy to Example 2.10 would not agree with φ at Lebesgue points in general. This illustrates the variety of extensions μ_x of δ_x provided by Proposition 2.13.

Proof of Proposition 2.13 For (1) it is sufficient to consider $m = 1$. We first notice that

$$\left| \int_\Omega \varphi \, d\mu \right| \le |\mu|(\Omega) \, \|\varphi\|_{C(\overline{\Omega})} \quad \text{for all} \quad \varphi \in C(\overline{\Omega}) .$$

Then $\varphi \to \int_\Omega \varphi \, d\mu$ is a linear continuous functional on $C(\overline{\Omega})$ and, by Riesz' Representation theorem, there is a Radon measure $\sigma \in \text{ca}(\overline{\Omega}, \mathcal{B}(\overline{\Omega}))$ such that

$$\int_{\overline{\Omega}} \varphi \, d\sigma = \int_\Omega \varphi \, d\mu \quad \text{for all} \quad \varphi \in C(\overline{\Omega}) .$$

For any $x \in \overline{\Omega} \setminus \text{core}\,\mu$ there is $\delta > 0$ with $B_\delta(x) \cap \text{core}\,\mu = \emptyset$, since $\text{core}\,\mu$ is closed. Thus,

$$\int_{\overline{\Omega}} \varphi \, d\sigma = \int_\Omega \varphi \, d\mu = 0$$

for all $\varphi \in C(\overline{\Omega})$ compactly supported on $B_\delta(x)$. Hence $\text{supp}\,\sigma \subset \text{core}\,\mu$.

For (2) we use the semi norm from (2.10) and observe that

$$\|\varphi\|_\Gamma = \|\varphi_{|\Gamma}\|_{C(\Gamma)} \le \|\varphi\|_{\mathcal{L}^\infty(\Omega)} \quad \text{for all} \quad \varphi \in C(\overline{\Omega}, \mathbb{R}^m) ,$$

since $|\varphi(x)| \leq \|\varphi\|_{\mathcal{L}^\infty(\Omega\cap B_\delta(x))}$ for all $x \in \Gamma$ and $\delta > 0$ by (2.11). Thus

$$\Big|\int_\Gamma \varphi\, d\sigma\Big| \leq |\sigma|(\Gamma)\, \|\varphi_{|\Gamma}\|_{C(\Gamma)} = |\sigma|(\Gamma)\, \|\varphi\|_\Gamma \quad \text{for all } \varphi \in C(\overline{\Omega}, \mathbb{R}^m)\,. \tag{2.14}$$

Hence $g_0^* : C(\overline{\Omega}, \mathbb{R}^m) \to \mathbb{R}$ with $\langle g_0^*, \varphi\rangle = \int_\Gamma \varphi\, d\sigma$ is a linear continuous functional on a subspace of $\mathcal{L}^\infty(\Omega, \mathbb{R}^m)$. By the Hahn-Banach theorem there is a linear continuous extension g^* of g_0^* to all of $\mathcal{L}^\infty(\Omega, \mathbb{R}^m)$ preserving (2.14). Thus

$$|\langle g^*, \varphi\rangle| \leq |\sigma|(\Gamma)\|\varphi\|_\Gamma \leq |\sigma|(\Gamma)\|\varphi\|_{\mathcal{L}^\infty(\Omega)} \quad \text{for all } \varphi \in \mathcal{L}^\infty(\Omega, \mathbb{R}^m)\,. \tag{2.15}$$

By Proposition 2.7 there is $\mu \in \mathrm{ba}(\Omega, \mathcal{B}(\Omega), \mathcal{L}^n)^m$ such that

$$\langle g^*, \varphi\rangle = \int_\Omega \varphi\, d\mu \quad \text{for all} \quad \varphi \in \mathcal{L}^\infty(\Omega, \mathbb{R}^m)\,.$$

Consequently,

$$|\mu|(\Omega) = \|g^*\| \leq |\sigma|(\Gamma)\,.$$

Since every $\varphi \in C_c(\Gamma, \mathbb{R}^m)$ can be extended to some function $\overline{\varphi} \in C(\overline{\Omega}, \mathbb{R}^m)$ under preservation of the norm (cf. [52, Proposition 2.1]) and since σ is bounded,

$$\begin{aligned} |\sigma|(\Gamma) = \sup_{\substack{\varphi\in C_c(\Gamma,\mathbb{R}^m)\\ \|\varphi\|_{C(\Gamma)}\leq 1}} \int_\Gamma \varphi\, d\sigma &= \sup_{\substack{\varphi\in C_c(\Gamma,\mathbb{R}^m)\\ \|\varphi\|_{C(\Gamma)}\leq 1}} \int_\Omega \overline{\varphi}\, d\mu \\ \leq \sup_{\substack{\psi\in C(\overline{\Omega},\mathbb{R}^m)\\ \|\psi\|_{C(\overline{\Omega})}\leq 1}} \int_\Omega \psi\, d\mu &= \sup_{\substack{\psi\in C(\overline{\Omega},\mathbb{R}^m)\\ \|\psi\|_{\mathcal{L}^\infty(\Omega)}\leq 1}} \int_\Omega \psi\, d\mu \leq |\mu|(\Omega)\,. \end{aligned}$$

Therefore $|\sigma|(\Gamma) = |\mu|(\Omega)$. If $\delta > 0$ and $\varphi \in \mathcal{L}^\infty(\Omega, \mathbb{R}^m)$ with $\varphi = 0$ on Γ_δ, then

$$\langle g^*, \varphi\rangle = \int_\Omega \varphi\, d\mu = 0$$

by (2.15). This implies $\mathrm{core}\,\mu \subset \overline{\Gamma}$. If $\mathcal{L}^n(\overline{\Gamma} \cap \Omega) = 0$, then μ is a pure measure by Proposition 2.12. □

For a measure $\mu \in \mathrm{ba}(\Omega, \mathcal{B}(\Omega), \mathcal{L}^n)$ with $\Omega \in \mathcal{B}(\mathbb{R}^n)$, all functions $f \in L^\infty(\Omega)$ are integrable by Proposition 2.7 and thus,

$$L^\infty(\Omega, \mathcal{B}(\Omega), \mathcal{L}^n) \subset L^1(\Omega, \mathcal{B}(\Omega), \mu)\,.$$

By a more detailed analysis of the special case $\mu = \mathrm{dens}_x^\Omega$ (cf. (2.13)) we not only show that the inclusion can be strict, but we also demonstrate how the integration theory can be applied to an important prototype of pure measures. We say that $f \in L^1_{\mathrm{loc}}(\Omega)$ has the *approximate limit* α at $x \in \overline{\Omega}$ (with respect to Ω), denoted by

$$\operatorname*{ap\,lim}_{y\to x} f(y) = \alpha\,,$$

if for all $\varepsilon > 0$

$$\lim_{\delta\to 0} \frac{\mathcal{L}^n\big(\Omega\cap B_\delta(x)\cap\{|f-\alpha|\geq\varepsilon\}\big)}{\mathcal{L}^n(\Omega\cap B_\delta(x))} = 0 \tag{2.16}$$

(cf. [22, p. 46]).

Proposition 2.15 *Let $\Omega \in \mathcal{B}(\mathbb{R}^n)$, let $f \in L^1_{\mathrm{loc}}(\overline{\Omega})$, let $x \in \overline{\Omega}$ be such that*

$$\mathcal{L}^n(\Omega\cap B_\delta(x)) > 0 \quad \textit{for all} \quad \delta > 0\,,$$

and let dens_x^Ω *be a density measure at x with respect to Ω.*

(1) *If there is some $\tilde{\delta} > 0$ such that*

$$\fint_{\Omega\cap B_\delta(x)} |f|\,d\mathcal{L}^n \quad \textit{is bounded for all} \quad 0<\delta<\tilde{\delta}\,,$$

then f is integrable with respect to dens_x^Ω *and*

$$\int_\Omega |f|\,d\,\mathrm{dens}_x^\Omega \leq \limsup_{\delta\to 0} \fint_{\Omega\cap B_\delta(x)} |f|\,d\mathcal{L}^n\,.$$

(2) *If we have*

$$\operatorname*{ap\,lim}_{y\to x} f(y) = \alpha \quad \textit{for some} \quad \alpha\in\mathbb{R}\,,$$

then $f = \alpha$ i.m. dens_x^Ω *and f is* dens_x^Ω*-integrable with*

$$\fint_{\{x\}} f\,d\,\mathrm{dens}_x^\Omega = \alpha\,.$$

(3) *If there is $\alpha \in \mathbb{R}$ with*

$$\lim_{\delta\to 0} \fint_{\Omega\cap B_\delta(x)} |f-\alpha|\,d\mathcal{L}^n = 0\,, \tag{2.17}$$

then $\operatorname{ap}\lim_{y\to x} f(y) = \alpha$ *and* f *is* $\operatorname{dens}_x^\Omega$*-integrable with*

$$\alpha = \fint_{\{x\}} f \, d \operatorname{dens}_x^\Omega = \lim_{\delta\to 0} \fint_{\Omega\cap B_\delta(x)} f \, d\mathcal{L}^n . \tag{2.18}$$

We first discuss the results and postpone the proof after Example 2.18.

Remark 2.16

(1) Let us discuss the results for Ω open and $x \in \Omega$. Notice first that, in this case, the results are also true for any density measure dens_x according to Example 2.4, since it agrees with some $\operatorname{dens}_x^\Omega$ on Ω. We also observe that (2.17) is valid with $\alpha = f(x)$ at any Lebesgue point $x \in \Omega$ of f by definition. This way we see that any $f \in L^1_{\mathrm{loc}}(\Omega)$ is $\operatorname{dens}_x^\Omega$-integrable at least for $\mathcal{L}^n$-a.e. $x \in \Omega$ (cf. [22, p. 44]). Consequently, for fixed $x \in \Omega$ there is a large class of $\operatorname{dens}_x^\Omega$-integrable functions beyond $L^\infty(\Omega)$. Notice that, conversely, $\operatorname{dens}_x^\Omega$-integrability of f does not imply that x is a Lebesgue point of f or that f has an approximate limit at x (take e.g. $f = \chi_A$ for some $A \subset \mathcal{B}(\Omega)$ such that liminf and limsup in (2.13) do not coincide).

(2) For Ω open the results also provide an integral representation for a slightly modified *precise representative* of f by

$$f^\times(x) := \begin{cases} \fint_{\{x\}} f \, d \operatorname{dens}_x^\Omega & \text{if the integral exists ,} \\ 0 & \text{otherwise .} \end{cases}$$

The commonly used precise representative, that equals

$$\lim_{\delta\to 0} \fint_{B_\delta(x)} f \, d\mathcal{L}^n$$

if the limit exists, obviously agrees with $f^\times(x)$ at all $x \in \Omega$ where (2.17) is satisfied. Thus it differs from $f^\times(x)$ at most on an $\mathcal{L}^n$-null set (cf. [22, p. 46]). More precisely, $f \in L^1_{\mathrm{loc}}(\Omega)$ might not be $\operatorname{dens}_x^\Omega$-integrable if merely $\lim_{\delta\to 0} \fint_{B_\delta(x)} f \, d\mathcal{L}^n$ exists, since roughly speaking $\fint_{\{x\}} f^\pm \, d \operatorname{dens}_x^\Omega$ can be infinite for the positive and negative part of f. If f is $\operatorname{dens}_x^\Omega$-integrable, the limit $\lim_{\delta\to 0} \fint_{B_\delta(x)} f \, d\mathcal{L}^n$ can exist but differ from $f^\times(x)$ (cf. Example 2.18 below). Moreover, f can be $\operatorname{dens}_x^\Omega$-integrable if the limit $\lim_{\delta\to 0} \fint_{B_\delta(x)} f \, d\mathcal{L}^n$ does not exist, since , e.g., the inequalities in (3.5) can be strict. In this last case the integral will depend on the special choice of $\operatorname{dens}_x^\Omega$.

For f in $W^{1,1}(\Omega)$ or $BV(\Omega)$ we readily obtain that it is integrable with respect to $\operatorname{dens}_x^\Omega$ for $\mathcal{H}^{n-1}$-a.e. $x \in \Omega$ and that $f^\times$ agrees with the usual precise representative $\mathcal{H}^{n-1}$-a.e. on Ω (cf. [22, pp. 160, 213]).

(3) If Ω is open with Lipschitz boundary and if f is in $W^{1,1}(\Omega)$ or $BV(\Omega)$, then the integral $\fint_{\{x\}} f \, d \operatorname{dens}_x^\Omega$ exists and agrees with the usual trace of f for $\mathcal{H}^{n-1}$-a.e.

$x \in \partial\Omega$ (cf. [22, pp. 133, 181]). Therefore, the precise representative $f^\times$ also provides an integral representation for the pointwise trace of f in this case.

(4) Let $\Omega \subset \mathbb{R}^n$ be open and let $f \in \mathcal{W}^{1,1}(\Omega, \mathbb{R}^m)$. For $\mathcal{L}^n$-a.e. $x \in \Omega$, the weak derivative $Df(x)$ agrees with the approximate derivative $D_{\text{ap}} f(x)$ defined by

$$\operatorname*{ap\,lim}_{y \to x} \frac{\big|f(y) - f(x) - D_{\text{ap}} f(x)(y-x)\big|}{|y-x|} = 0$$

(cf. [22, p. 233]). Thus, by Proposition 2.15,

$$⨍_{\{x\}} \frac{\big|f(y) - f(x) - Df(x)(y-x)\big|}{|y-x|}\, d\,\text{dens}_x^\Omega = 0 \quad \text{for} \quad \mathcal{L}^n\text{-a.e. } x \in \Omega\,.$$

For $y = x + th$ with $h \in \mathbb{R}^n$ and $t \in \mathbb{R}$, we get

$$Df(x)\,h = \operatorname*{ap\,lim}_{t \to 0} \frac{f(x+th) - f(x)}{t}\,.$$

With a density measure $\text{dens}_0^{\mathbb{R}} \in \text{ba}(\mathbb{R}, \mathcal{B}(\mathbb{R}), \mathcal{L}^1)$ at $t = 0$ in $\mathbb{R}$, this gives

$$Df(x)\,h = ⨍_{\{0\}} \frac{f(x+th) - f(x)}{t}\, d\,\text{dens}_0^{\mathbb{R}} \quad \text{for all } h \in \mathbb{R}^n\,,\ \mathcal{L}^n\text{-a.e. } x \in \Omega\,,$$

which is an integral representation for weak directional derivatives.

For $f \in BV(\Omega)$ we have an even stronger result saying that, on a jump set of f, the precise representative $f^\times$ gives the mean value of the approximate limits from both sides for $\mathcal{H}^{n-1}$-a.e. x (cf. also [22, p. 213], [2, p. 175]).

Corollary 2.17 *Let $\Omega \subset \mathbb{R}^n$ be open, let $f \in BV(\Omega)$, and let* dens_x^Ω *be any density measure at $x \in \Omega$ with respect to Ω. Then*

$$f^\times(x) = ⨍_{\{x\}} f\, d\,\text{dens}_x^\Omega = \frac{f^i(x) + f^s(x)}{2} \qquad \mathcal{H}^{n-1}\text{-a.e. on } \Omega$$

where

$$f^i(x) := \operatorname*{ap\,lim\,inf}_{y \to x} f(y) := \sup\left\{\alpha \;\middle|\; \lim_{\delta \to 0} \frac{\mathcal{L}^n\big(B_\delta(x) \cap \{f < \alpha\}\big)}{\mathcal{L}^n(B_\delta(x))} = 0\right\},$$

$$f^s(x) := \operatorname*{ap\,lim\,sup}_{y \to x} f(y) := \inf\left\{\alpha \;\middle|\; \lim_{\delta \to 0} \frac{\mathcal{L}^n\big(B_\delta(x) \cap \{f > \alpha\}\big)}{\mathcal{L}^n(B_\delta(x))} = 0\right\}$$

are the lower *and the* upper approximate limit *of f at x, respectively.*

Let us discuss an estimate before we give the proof at the end of this section.

For $\Omega \in \mathcal{B}(\mathbb{R}^n)$ and $x \in \overline{\Omega}$ such that $\mathcal{L}^n(\Omega \cap B_\delta(x)) > 0$ for all $\delta > 0$, we readily get from (2.13) for all $A \in \mathcal{B}(\Omega)$

$$\int_\Omega \chi_A \, d\,\mathrm{dens}_x^\Omega = \mathrm{dens}_x^\Omega(A) \le \limsup_{\delta\to 0} \frac{\mathcal{L}^n\big(A \cap B_\delta^\Omega(x)\big)}{\mathcal{L}^n(B_\delta^\Omega(x))} = \limsup_{\delta\to 0} \fint_{B_\delta^\Omega(x)} \chi_A \, d\mathcal{L}^n$$

and an analogous relation with $\liminf$. This implies for simple functions h that

$$\liminf_{\delta\to 0} \fint_{B_\delta^\Omega(x)} h \, d\mathcal{L}^n \le \int_\Omega h \, d\,\mathrm{dens}_x^\Omega \le \limsup_{\delta\to 0} \fint_{B_\delta^\Omega(x)} h \, d\mathcal{L}^n . \tag{2.19}$$

Now we can ask to what extent this remains true for dens_x^Ω-integrable $f \in L^1_{\mathrm{loc}}(\Omega)$. For any $f \in L^\infty(\Omega)$ this follows by uniform approximation with simple functions. If $f \in L^1_{\mathrm{loc}}(\Omega)$ satisfies (2.17), then (2.19) follows with equality. Let us provide an example that this might fail if merely $\mathrm{ap\,lim}_{y\to x} f(y) = \alpha$.

Example 2.18 Let $\Omega = B_1(0) \subset \mathbb{R}^2$. With polar coordinates (r, β) we set (radius $r > 0$, angle $\beta \in [0, 2\pi)$)

$$\Omega' = \big\{(r, \beta) \,\big|\, 0 < r < 1,\ 0 < \beta < r\big\} \subset \Omega .$$

Now we consider $f \in L^1(\Omega)$ given by

$$f(x) = \begin{cases} \frac{1}{|x|} & \text{on } \Omega' , \\ 0 & \text{otherwise} . \end{cases}$$

Then, for $0 < \delta < 1$,

$$\fint_{B_\delta(0)} f \, d\mathcal{L}^2 = \frac{1}{\pi\delta^2} \int_0^\delta \int_0^r \frac{1}{r} r \, d\beta \, dr = \frac{1}{\pi\delta^2} \int_0^\delta r \, dr = \frac{1}{2\pi}$$

From (2.13) we obtain for $x = 0$

$$0 \le \mathrm{dens}_0^\Omega(\Omega') \le \limsup_{\delta\downarrow 0} \frac{\mathcal{L}^2(\Omega' \cap B_\delta(0))}{\mathcal{L}^2(B_\delta(0))} = 0 .$$

Since $\Omega' \cap B_{\frac{1}{\varepsilon}}(0) = \{|f - 0| > \varepsilon\}$ for $\varepsilon > 0$,

$$\mathop{\mathrm{ap\,lim}}_{y\to 0} f(y) = 0 . \tag{2.20}$$

Consequently, by Proposition 2.15, $f = 0$ i.m. dens_0^Ω and

$$0 = \fint_{\{0\}} f \, d\,\mathrm{dens}_0^\Omega < \liminf_{\delta \to 0} \fint_{B_\delta(0)} f \, d\mathcal{L}^2 . \tag{2.21}$$

For $-f$ we get the opposite inequality with $\limsup$. Thus (2.19) is not valid with $\pm f$ instead of h. Since $\lim_{\delta \to 0} \fint_{B_\delta(0)} f \, d\mathcal{L}^2 > 0$ exists, $f^\times(x) = 0$ differs from the usual precise representative. If we define f with $\frac{1}{|x|^2}$ instead of $\frac{1}{|x|}$ on Ω', then we still have (2.20) but

$$\fint_{B_\delta(0)} f \, d\mathcal{L}^2 = \frac{1}{\pi \delta} \overset{\delta \to 0}{\longrightarrow} \infty ,$$

i.e. the equality in (2.21) remains the same while the $\liminf$ becomes even infinite. The example also shows that the boundedness of $\fint_{\Omega \cap B_\delta(0)} |f| \, d\mathcal{L}^n$ for small $\delta > 0$ is not sufficient for the second equality in (2.18).

Proof of Proposition 2.15 We have that $f \in L^1(\Omega \cap B_{\bar\delta}(x))$ for some $\bar\delta > 0$.

For (1) there is some $c > 0$ such that $\fint_{B_\delta^\Omega(x)} |f| \, d\mathcal{L}^n < c$ for all $\delta \in (0, \tilde\delta)$. We define for each $k \in \mathbb{N}$

$$\Omega_k := \big\{ y \in \Omega \; \big| \; |f(y)| < k \big\} , \quad \Omega_k^0 := \Omega \setminus \Omega_k ,$$

$$h_k(x) := \begin{cases} \frac{l}{2^k} & \text{on } |f|^{-1}\big(\big[\frac{l}{2^k}, \frac{l+1}{2^k}\big)\big) \cap \Omega_k \text{ for all } l \in \mathbb{N} , \\ 0 & \text{on } \Omega_k^0 . \end{cases}$$

Obviously all h_k are Borel measurable and thus, they are simple functions related to dens_x^Ω. Clearly, $h_k \le |f|$ on Ω for all $k \in \mathbb{N}$. By $|\,h_k - |f|\,| < \frac{1}{2^k}$ on Ω_k we get

$$\big\{ y \in \Omega \; \big| \; |\,h_k - |f|\,| > \varepsilon \big\} \subset \Omega_k^0 \quad \text{if} \quad \tfrac{1}{2^k} < \varepsilon .$$

Therefore

$$\mathrm{dens}_x^\Omega\big(\big\{ |\,h_k - |f|\,| > \varepsilon \big\}\big) \le \mathrm{dens}_x^\Omega(\Omega_k^0) . \tag{2.22}$$

Let us first assume that $\mathrm{dens}_x^\Omega(\Omega_k^0) \ge \frac{2}{\sqrt{k}}$. Then we choose $\delta_k \in (0, \tilde\delta)$ such that

$$\limsup_{\delta \to 0} \frac{\mathcal{L}^n\big(\Omega_k^0 \cap B_\delta^\Omega(x)\big)}{\mathcal{L}^n(B_\delta^\Omega(x))} \le \frac{\mathcal{L}^n\big(\Omega_k^0 \cap B_{\delta_k}^\Omega(x)\big)}{\mathcal{L}^n(B_{\delta_k}^\Omega(x))} + \frac{1}{\sqrt{k}} .$$

Consequently,

$$
\begin{aligned}
&\fint_{B^{\Omega}_{\delta_k}(x)} |f|\, d\mathcal{L}^n \\
&\quad = \frac{1}{\mathcal{L}^n(B^{\Omega}_{\delta_k}(x))}\left(\int_{B^{\Omega}_{\delta_k}(x)\cap\Omega_k} |f|\, d\mathcal{L}^n + \int_{B^{\Omega}_{\delta_k}(x)\cap\Omega_k^0} |f|\, d\mathcal{L}^n\right) \\
&\quad \ge k\, \frac{\mathcal{L}^n\big(\Omega_k^0 \cap B^{\Omega}_{\delta_k}(x)\big)}{\mathcal{L}^n(B^{\Omega}_{\delta_k}(x))} \\
&\quad \ge k\left(\limsup_{\delta\to 0} \frac{\mathcal{L}^n\big(\Omega_k^0 \cap B^{\Omega}_{\delta}(x)\big)}{\mathcal{L}^n(B^{\Omega}_{\delta}(x))} - \frac{1}{\sqrt{k}}\right) \\
&\quad \overset{(2.13)}{\ge} k\left(\mathrm{dens}^{\Omega}_x(\Omega_k^0) - \tfrac{1}{\sqrt{k}}\right) \ge \sqrt{k}\,.
\end{aligned}
$$

But this is impossible for $\sqrt{k} > c$ by the boundedness of the left hand side and, therefore, $\mathrm{dens}^{\Omega}_x(\Omega_k^0) < \frac{2}{\sqrt{k}}$ for such k. Using (2.22) we get

$$
h_k \xrightarrow{\mathrm{dens}^{\Omega}_x} |f|\,.
$$

With (2.13) we obtain for all $A \in \mathcal{B}(\Omega)$

$$
\begin{aligned}
\int_{\Omega} \chi_A\, d\,\mathrm{dens}^{\Omega}_x = \mathrm{dens}^{\Omega}_x(A) &\le \limsup_{\delta\to 0} \frac{\mathcal{L}^n\big(A \cap B^{\Omega}_{\delta}(x)\big)}{\mathcal{L}^n(B^{\Omega}_{\delta}(x))} \\
&= \limsup_{\delta\to 0} \fint_{B^{\Omega}_{\delta}(x)} \chi_A\, d\mathcal{L}^n
\end{aligned}
$$

and an analogous relation with $\liminf$. This gives for the simple functions h_k

$$
\liminf_{\delta\to 0} \fint_{B^{\Omega}_{\delta}(x)} h_k\, d\mathcal{L}^n \le \int_{\Omega} h_k\, d\,\mathrm{dens}^{\Omega}_x \le \limsup_{\delta\to 0} \fint_{B^{\Omega}_{\delta}(x)} h_k\, d\mathcal{L}^n\,.
$$

Since $0 \le h_k \le |f|$ on Ω, we obtain that

$$
0 \le \int_{\Omega} h_k\, d\,\mathrm{dens}^{\Omega}_x \le \limsup_{\delta\to 0} \fint_{B^{\Omega}_{\delta}(x)} |f|\, d\mathcal{L}^n < c\,. \tag{2.23}
$$

By construction, the sequence $\{h_k\}$ of simple functions is increasing. Hence

$$
\int_{\Omega} |h_k - h_l|\, d\,\mathrm{dens}^{\Omega}_x \to 0 \quad \text{as} \quad k, l \to \infty\,.
$$

Consequently, $|f|$ is dens_x^Ω-integrable with determining sequence $\{h_k\}$ and hence also f is dens_x^Ω-integrable. Taking the limit $k \to \infty$ in (2.23) we get the remaining estimate.

For (2) we fix $\varepsilon > 0$ and set $\Omega^\varepsilon := \big\{ y \in \Omega \,\big|\, |f - \alpha| \geq \varepsilon \big\}$. Then, by (2.13),

$$0 \leq \mathrm{dens}_x^\Omega(\Omega^\varepsilon) \leq \limsup_{\delta \to 0} \frac{\mathcal{L}^n\big(\Omega^\varepsilon \cap B_\delta^\Omega(x)\big)}{\mathcal{L}^n(B_\delta^\Omega(x))} = 0$$

where the last equality follows from $\operatorname{ap\,lim}_{y\to x} f(y) = \alpha$. Thus, $\mathrm{dens}_x^\Omega(\Omega^\varepsilon) = 0$ for all $\varepsilon > 0$, which implies $f = \alpha$ i.m. on Ω. Hence the constant sequence $\{\alpha\}_k$ is a determining sequence for f. Therefore f is dens_x^Ω-integrable with

$$\fint_{\{x\}} f \, d\,\mathrm{dens}_x^\Omega = \int_\Omega \alpha \, d\,\mathrm{dens}_x^\Omega = \alpha\, \mathrm{dens}_x^\Omega(\Omega) = \alpha\,.$$

For (3) we use Ω^ε as defined in the proof of (2) to get

$$\begin{aligned} 0 &\leq \limsup_{\delta\to 0} \frac{\mathcal{L}^n\big(\Omega^\varepsilon \cap B_\delta^\Omega(x)\big)}{\mathcal{L}^n(B_\delta^\Omega(x))} \\ &\leq \limsup_{\delta\to 0} \frac{1}{\varepsilon \mathcal{L}^n(B_\delta^\Omega(x))} \int_{\Omega^\varepsilon \cap B_\delta^\Omega(x)} |f - \alpha| \, d\mathcal{L}^n \\ &\leq \limsup_{\delta\to 0} \frac{1}{\varepsilon} \fint_{B_\delta^\Omega(x)} |f - \alpha| \, d\mathcal{L}^n \;=\; 0\,. \end{aligned}$$

Hence $\operatorname{ap\,lim}_{y\to x} f(y) = \alpha$ by the definition of Ω^ε. Thus (2) implies that f is dens_x^Ω-integrable with $\fint_{\{x\}} f \, d\,\mathrm{dens}_x^\Omega = \alpha$. By (2.17),

$$\Big| \fint_{\Omega\cap B_\delta(x)} \alpha - f \; d\mathcal{L}^n \Big| \leq \fint_{\Omega\cap B_\delta(x)} |\alpha - f| \, d\mathcal{L}^n \overset{\delta\to 0}{\longrightarrow} 0\,.$$

Hence

$$\alpha = \lim_{\delta\to 0} \fint_{\Omega\cap B_\delta(x)} \alpha \, d\mathcal{L}^n = \lim_{\delta\to 0} \fint_{\Omega\cap B_\delta(x)} f \, d\mathcal{L}^n$$

which verifies the assertion. □

Proof of Corollary 2.17 We have

$$-\infty < f^i(x) \leq f^s(x) < \infty \quad \text{for } \mathcal{H}^{n-1}\text{-a.e. } x \in \Omega \tag{2.24}$$

(cf. [22, p. 211]). We use $\tilde{f}(x) = \frac{1}{2}(f^i(x) + f^s(x))$ and show the statement for $x \in \Omega$ where (2.24) is satisfied. If the approximate limit of f at x exists, then

$$f^i(x) = f^s(x) = \operatorname*{ap\,lim}_{y \to x} f(y)$$

and $f^\times(x) = \tilde{f}(x)$ by Proposition 2.15. Otherwise there are disjoint open half spaces $H_\pm \subset \mathbb{R}^n$ such that $x \in \partial H_\pm$ and

$$f^i(x) = \operatorname*{ap\,lim}_{\substack{y \to x \\ y \in H_-}} f(y), \qquad f^s(x) = \operatorname*{ap\,lim}_{\substack{y \to x \\ y \in H_+}} f(y)$$

(cf. [22, p. 213] and notice that merely half balls $B_\delta^{H_\pm}(x)$ enter the computation of ap lim). By $\mathcal{L}^n(H^+ \cap H^-) = 0$, the measures

$$\mathrm{dens}_x^{\Omega \cap H^\pm} := 2\,\mathrm{dens}_x^\Omega \llcorner H^\pm$$

are density measures at x with respect to $H^\pm$. From Proposition 2.15 (2) with $\Omega \cap H^\pm$ instead of Ω we get

$$f^i(x) = \fint_{\{x\}} f \, d\,\mathrm{dens}_x^{\Omega \cap H^-}, \qquad f^s(x) = \fint_{\{x\}} f \, d\,\mathrm{dens}_x^{\Omega \cap H^+}.$$

With

$$\mathrm{dens}_x^\Omega = \tfrac{1}{2}\big(\mathrm{dens}_x^{\Omega \cap H^-} + \mathrm{dens}_x^{\Omega \cap H^-}\big)$$

we get $f^\times(x) = \tilde{f}(x)$ also in this case. $\square$

Chapter 3
Theory of Traces

For the treatment of partial differential equations, Sobolev and BV functions play an essential role. Since they cannot be evaluated directly on the boundary, it is common to consider a trace operator for sufficiently regular Ω. As typical example available today we can take $\Omega \subset \mathbb{R}^n$ open and bounded with Lipschitz boundary to have a linear continuous operator

$$T : \mathcal{W}^{1,1}(\Omega) \to \mathcal{L}^1(\partial\Omega, \mathcal{H}^{n-1})$$

such that for all $f \in \mathcal{W}^{1,1}(\Omega)$ and $\varphi \in C^1(\overline{\Omega}, \mathbb{R}^n)$

$$\int_\Omega f \operatorname{div}\varphi \, d\mathcal{L}^n + \int_\Omega \varphi \cdot Df \, d\mathcal{L}^n = \int_{\partial\Omega} \varphi \cdot \nu^\Omega \, Tf \, d\mathcal{H}^{n-1} \tag{3.1}$$

(cf. [22, p. 133], [32, p. 168]). Here the surface integral on the right hand side is related to the vector-valued Radon measure $\nu^\Omega \, Tf \mathcal{H}^{n-1} \lfloor \partial\Omega$. This basically restricts (3.1) to sets Ω of finite perimeter, since these are the sets having a suitable normal field on their boundary. We will overcome that limitation by a much more general approach.

3.1 General Traces

Notice that the left hand side in (3.1) can be considered as linear continuous functional $f^* \in C^1(\overline{\Omega}, \mathbb{R}^n)^*$ such that

$$\langle f^*, \varphi \rangle = 0 \quad \text{if} \quad \varphi_{|\partial\Omega} = 0 \, . \tag{3.2}$$

With this observation in mind we introduce a more general notion of trace. Let $U \subset \mathbb{R}^n$ be a Borel set, let $\Gamma \subset \overline{U}$, and let X be a normed space of functions

F. Schuricht, M. Schönherr, *A Theory of Traces and the Divergence Theorem*, Lecture Notes in Mathematics 2372, https://doi.org/10.1007/978-3-031-86664-7_3

$\varphi : U \to \mathbb{R}^m$. A *trace* or *trace functional* on Γ over X is some $f^* \in X^*$ such that for all $\varphi \in X$

$$\langle f^*, \varphi \rangle = 0 \qquad \text{if} \qquad \varphi_{|\Gamma_\delta \cap U} = 0 \;\text{ for some }\; \delta > 0\,. \tag{3.3}$$

Clearly f^* in (3.2) is a trace on $\partial\Omega$ related to f. Since $\Gamma_\delta = \big(\overline{\Gamma}\big)_\delta$ for all $\delta > 0$, it is sufficient to consider traces on closed Γ.

Let us motivate our approach by some traces over $\mathcal{L}^\infty(U)$ which show in particular that (3.2) is too restrictive for general traces which we intend to study.

Proposition 3.1 *Let $U \subset \mathbb{R}^n$ be a Borel set, let $E \in \mathcal{B}(U)$, let $\Gamma \subset \overline{U}$ be closed, and let $\gamma : (0, \infty) \to (0, \infty)$ be continuous such that*

$$c := \limsup_{\delta \downarrow 0} \frac{1}{\gamma(\delta)} \int_{\Gamma_\delta \cap E} d\mathcal{L}^n \quad \textit{is finite.} \tag{3.4}$$

Then there exists a measure

$$\mu_\Gamma \in \mathrm{ba}\big(U, \mathcal{B}(U), \mathcal{L}^n\big) \quad \textit{with} \quad \operatorname{core} \mu_\Gamma \subset \Gamma\,, \quad |\mu_\Gamma|(U) \le c$$

such that $f^*_\Gamma \in \mathcal{L}^\infty(U)^*$ related to $f\mu_\Gamma$ is a trace on Γ for all $f \in \mathcal{L}^\infty(U)$ and

$$\begin{aligned} \liminf_{\delta \downarrow 0} \frac{1}{\gamma(\delta)} \int_{\Gamma_\delta \cap E} \varphi f \, d\mathcal{L}^n \le \langle f^*_\Gamma, \varphi \rangle &= \int_\Gamma \varphi f \, d\mu_\Gamma \\ &\le \limsup_{\delta \downarrow 0} \frac{1}{\gamma(\delta)} \int_{\Gamma_\delta \cap E} \varphi f \, d\mathcal{L}^n \end{aligned} \tag{3.5}$$

for all $\varphi \in \mathcal{L}^\infty(U)$. If the limsup in (3.4) is a limit, then $|\mu_\Gamma|(U) = c$. The mapping

$$T : \mathcal{L}^\infty(U) \to \mathcal{L}^\infty(U)^* \quad \textit{with} \quad Tf = f^*_\Gamma$$

is linear and continuous. For fixed $f, \varphi \in \mathcal{L}^\infty(U)$ there is a sequence $\delta_j \downarrow 0$ with

$$\int_\Gamma \varphi f \, d\mu_\Gamma = \lim_{j \to \infty} \frac{1}{\gamma(\delta_j)} \int_{\Gamma_{\delta_j} \cap E} \varphi f \, d\mathcal{L}^n\,. \tag{3.6}$$

Let us discuss the result before we give the proof at the end of this section. For a nontrivial measure μ_Γ one obviously needs

$$\int_{\Gamma_\delta \cap E} d\mathcal{L}^n = \mathcal{L}^n(\Gamma_\delta \cap E) > 0 \quad \text{for all } \delta > 0\,.$$

In applications we consider the special choices $\gamma(\delta) = \delta$ and

$$\gamma(\delta) = \mathcal{L}^n(\Gamma_\delta \cap E) \quad \left(\text{thus } \frac{1}{\gamma(\delta)} \int_{\Gamma_\delta \cap E} \varphi f \, d\mathcal{L}^n = \fint_{\Gamma_\delta \cap E} \varphi f \, d\mathcal{L}^n \right). \tag{3.7}$$

Notice that $|\mu_\Gamma|(U) = 1$ in case (3.7). In some examples the following approximation result turns out to be helpful.

Corollary 3.2 *Let $U \subset \mathcal{B}(\mathbb{R}^n)$, $E \in \mathcal{B}(U)$, $\Gamma \subset \overline{U}$, $\gamma \in C(\mathbb{R}_{>0})$, and μ_Γ be as in Proposition* 3.1. *Moreover let $\tilde{\delta} > 0$ and $\varphi_k, \varphi \in \mathcal{L}^\infty(U)$ be such that*

$$\frac{1}{\gamma(\delta)} \int_{\Gamma_\delta \cap E} \varphi \, d\mathcal{L}^n = \lim_{k \to \infty} \frac{1}{\gamma(\delta)} \int_{\Gamma_\delta \cap E} \varphi_k \, d\mathcal{L}^n \quad \textit{uniformly for } \delta \in (0, \tilde{\delta}), \tag{3.8}$$

$$\fint_\Gamma \varphi_k \, d\mu_\Gamma = \lim_{\delta \downarrow 0} \frac{1}{\gamma(\delta)} \int_{\Gamma_\delta \cap E} \varphi_k \, d\mathcal{L}^n \quad \textit{for all } k \in \mathbb{N}. \tag{3.9}$$

Then

$$\fint_\Gamma \varphi \, d\mu_\Gamma = \lim_{k \to \infty} \fint_\Gamma \varphi_k \, d\mu_\Gamma .$$

Let us also postpone the proof of the corollary to the end of this section and continue with the discussion of the results.

For $E = U$, $\Gamma = \{x\}$ and γ as in (3.7), Proposition 3.1 provides a measure μ_x, that we call dens_x^U in accordance with (2.13), and a trace f_x^* on $\{x\}$ over $\mathcal{L}^\infty(U)$. We readily notice that f_x^* cannot satisfy (3.2) with $\{x\}$ instead of $\partial\Omega$, since $\varphi_{|\{x\}}$ cannot be defined in a reasonable way. However, the trace f_x^* provides an evaluation of f at x due to

$$\bar{f}(x) := \fint_{\{x\}} f \, d\,\mathrm{dens}_x^U = \langle f_x^*, 1 \rangle \quad \text{for all} \quad f \in \mathcal{L}^\infty(\Omega) .$$

By (3.5) this agrees with $f(x)$ if x is Lebesgue point of $f \in L^\infty(U)$ (let us mention that this needs not be the case for any extension μ_x of δ_x according to Proposition 2.13 as, e.g., in Example 2.14). The mapping $f \to f_x^*$ can be considered as a trace operator on $\mathcal{L}^\infty(\Omega)$ at x. If we fix f and vary x, then we have a pointwise integral representation of f that agrees a.e. with its precise representative $f^\times$ (cf. Remark 2.16).

Let us discuss Proposition 3.1 with

$$U = E = (0, 1)^2 \subset \mathbb{R}^2 , \quad \Gamma = \partial U , \quad \gamma(\delta) = \delta .$$

Then, for fixed $f \in \mathcal{L}^\infty(U)$, the functional $f^*_{\partial U}$ is a trace on ∂U. For $g, \varphi \in \mathcal{L}^\infty(U)$ and the related $\mu_{\partial U}$, we obviously have

$$\left| \fint_{\partial U} \varphi(f-g)\, d\mu_{\partial U} \right| \le \|\varphi\|_{\mathcal{L}^\infty(U)} \, \|f-g\|_{\mathcal{L}^\infty((\partial U)_\delta)} \quad \text{for all} \quad \delta > 0 \,.$$

Therefore the trace $g^*_{\partial U} = g\mu_{\partial U}$ agrees with $f^*_{\partial U}$ for all functions g in the affine linear subspace

$$X_f = \left\{ g \in \mathcal{L}^\infty(U) \,\middle|\, \|f-g\|_{\partial U} = 0 \right\}$$

(cf. (2.10)). This somehow means that $g \in X_f$ behaves as f arbitrarily close to ∂U and $f^*_{\partial U}$ appears to be an appropriate tool to describe that behavior. If we restrict our attention to $\varphi \in C(\overline{U})$, then we can identify $f^*_{\partial U}$ with a σ-measure $f^\sigma_{\partial U}$ supported on ∂U. By the application of (3.6) to mollified versions of $\varphi = \chi_R$ with rectangles R intersecting ∂U, we obtain that $f^\sigma_{\partial U}$ is (weakly) absolutely continuous with respect to $\mathcal{H}^1 \lfloor \partial U$. Hence there is a density function f^σ on ∂U such that $f^\sigma_{\partial U} = f^\sigma \mathcal{H}^1 \lfloor \partial U$. However, in the general case with $\varphi \in \mathcal{L}^\infty(U)$ we cannot find a function on ∂U representing the measure $f^*_{\partial U}$.

We now consider

$$\Omega = \Omega_0 \cup \Omega_1 \subset \mathbb{R}^2 \quad \text{with} \quad \Omega_j = (j, j+1) \times (0,1)$$

and $f \in \mathcal{L}^\infty(\Omega)$ given by

$$f = 0 \;\text{ on }\; \Omega_0 \,, \qquad f = 1 \;\text{ on }\; \Omega_1$$

(which is merely a representative of an equivalence class). For $\Gamma = \partial\Omega$ and $\gamma(\delta) = \delta$ Proposition 3.1 provides a measure μ_Γ and a trace f^*_Γ on Γ such that

$$\langle f^*_\Gamma, \varphi \rangle = \fint_\Gamma \varphi f \, d\mu_\Gamma \quad \text{for all} \quad \varphi \in \mathcal{L}^\infty(\Omega) \,.$$

In the light of usual traces we can try to assign a function f_Γ on Γ to the trace f^*_Γ by the requirement

$$\fint_\Gamma \varphi f \, d\mu_\Gamma = \int_\Gamma \varphi f_\Gamma \, d\mathcal{H}^1 \tag{3.10}$$

for suitable φ that, however, have to be extendable up to Γ. For $\varphi \in C(\overline{\Omega})$ we get from (3.5)

$$f_\Gamma = \begin{cases} 0 & \text{on } \partial\Omega_0 \setminus \partial\Omega_1\,, \\ 1 & \text{on } \partial\Omega_1 \setminus \partial\Omega_0\,, \\ \frac{1}{2} & \text{on } \partial\Omega_0 \cap \partial\Omega_1\,. \end{cases}$$

But notice that f_Γ cannot provide the precise behavior of f near $\partial\Omega_0 \cap \partial\Omega_1$ while f^*_Γ can give the full information by using $\varphi \in C(\Omega)$. In addition, f^*_Γ is not restricted to φ that are extendable up to Γ. This property of general (finitely additive) measures allows the construction of much more general Gauss-Green formulas than before.

Finally let us roughly sketch how we extend the Gauss-Green formula (3.1) to general Borel sets Ω contained in some open set $U \subset \mathbb{R}^n$. For $f \in \mathcal{W}^{1,1}(U)$ the left hand side can obviously be considered as functional $f^* \in \mathcal{W}^{1,\infty}(U, \mathbb{R}^n)^*$. Based on Proposition 2.7 it can be shown that f^* is related to measures

$$\lambda_f \in \mathrm{ba}\big(U, \mathcal{B}(U), \mathcal{L}^n\big)^n \quad \text{and} \quad \mu_f \in \mathrm{ba}\big(U, \mathcal{B}(U), \mathcal{L}^n\big)^{n\times n}$$

such that

$$\int_\Omega f \operatorname{div}\varphi \, d\mathcal{L}^n + \int_\Omega \varphi \cdot Df \, d\mathcal{L}^n = \langle \lambda_f, \varphi \rangle + \langle \mu_f, D\varphi \rangle$$

for all $\varphi \in \mathcal{W}^{1,\infty}(U, \mathbb{R}^n)$ where, in full generality, the core of λ_f and μ_f belongs to a small neighborhood of $\partial\Omega$. In 'better' cases their core belongs to $\partial\Omega$ and we get

$$\int_\Omega f \operatorname{div}\varphi \, d\mathcal{L}^n + \int_\Omega \varphi \cdot Df \, d\mathcal{L}^n = \fint_{\partial\Omega} \varphi \, d\lambda_f + \fint_{\partial\Omega} D\varphi \, d\mu_f\,.$$

In 'even better' cases λ_f can be considered as Radon measure on $\partial\Omega$ and μ_f might disappear. If Ω has some inner boundary, the measures 'know' the function f on both sides of it and λ_f cannot be a σ-measure as in the previous example surrounding (3.10). In some cases, the measure μ_f disappears and we get more structure for the other boundary term where, in particular, f enters explicitly. More precisely, in these cases there is a so-called normal measure $\nu \in \mathrm{ba}\big(U, \mathcal{B}(U), \mathcal{L}^n\big)^n$ with $\operatorname{core}\nu \subset \partial\Omega$, an extension of the pointwise outer normal function, such that

$$\int_\Omega f \operatorname{div}\varphi \, d\mathcal{L}^n + \int_\Omega \varphi \cdot Df \, d\mathcal{L}^n = \fint_{\partial\Omega} f\varphi \, d\nu \quad \text{for all } \varphi \in \mathcal{W}^{1,\infty}(\Omega, \mathbb{R}^n)\,.$$

For some normal measures ν we even get that

$$\nu = \nu^\Omega \omega_{\partial\Omega}$$

for the normal field $\nu^\Omega = D\big(\mathrm{dist}_\Omega - \mathrm{dist}_{\Omega^c}\big)$ on Ω and some measure $\omega_{\partial\Omega}$ near the boundary $\partial\Omega$ as in Proposition 3.1.

Summarizing it turns out that the results derived below would not be possible in that generality with a notion of trace relying merely on pointwise trace functions on the boundary $\partial\Omega$. We will develop our theory first for vector fields having divergence measure. Then the results for Sobolev functions and BV functions are more or less direct consequences.

Proof of Proposition 3.1 Let $X := \mathcal{L}^\infty(U)$. Then

$$X_0 := \left\{\varphi \in X \;\middle|\; \lim_{\delta\downarrow 0} \frac{1}{\gamma(\delta)} \int_{\Gamma_\delta \cap E} \varphi \, d\mathcal{L}^n \text{ exists}\right\}$$

is a linear subspace. $g_0^* : X_0 \to \mathbb{R}$ with

$$g_0^*(\varphi) := \lim_{\delta\downarrow 0} \frac{1}{\gamma(\delta)} \int_{\Gamma_\delta \cap E} \varphi \, d\mathcal{L}^n$$

is a continuous linear functional on X_0 majorized by the positively homogeneous and subadditive functional $\tilde{g} : \mathcal{L}^\infty(U) \to \mathbb{R}$ given by

$$\tilde{g}(\varphi) := \limsup_{\delta\downarrow 0} \frac{1}{\gamma(\delta)} \int_{\Gamma_\delta \cap E} \varphi \, d\mathcal{L}^n \le c\|\varphi\|_{\mathcal{L}^\infty} .$$

The Hahn-Banach theorem provides an extension $g^* \in \mathcal{L}^\infty(U)^*$ of g_0^* that is also majorized by $\tilde{g}$ on X. Hence $\|g^*\| \le c$ and

$$\begin{aligned} \liminf_{\delta\downarrow 0} \frac{1}{\gamma(\delta)} \int_{\Gamma_\delta \cap E} \varphi \, d\mathcal{L}^n &= -\limsup_{\delta\downarrow 0} \frac{1}{\gamma(\delta)} \int_{\Gamma_\delta \cap E} -\varphi \, d\mathcal{L}^n = -\tilde{g}(-\varphi) \\ &\le -\langle g^*, -\varphi\rangle = \langle g^*, \varphi\rangle \\ &\le \tilde{g}(\varphi) = \limsup_{\delta\downarrow 0} \frac{1}{\gamma(\delta)} \int_{\Gamma_\delta \cap E} \varphi \, d\mathcal{L}^n \end{aligned} \tag{3.11}$$

for all $\varphi \in \mathcal{L}^\infty(U)$. In the case where the limsup in (3.4) is a limit, we have for $\varphi \equiv 1$ that $\|\varphi\|_{\mathcal{L}^\infty} = 1$ and $\langle g^*, \varphi\rangle = c$ by (3.11) and thus, $\|g^*\| = c$. If $\varphi_{|\Gamma_{\delta'}} = 0$ for some $\delta' > 0$, then obviously

$$0 = \lim_{\delta\downarrow 0} \frac{1}{\gamma(\delta)} \int_{\Gamma_\delta \cap E} \varphi \, d\mathcal{L}^n = \langle g^*, \varphi\rangle \tag{3.12}$$

and hence g^* is a trace on Γ.

Let $\mu_\Gamma \in \mathrm{ba}\big(U, \mathcal{B}(U), \mathcal{L}^n\big)$ be related to $g^* \in \mathcal{L}^\infty(U)^*$ as in Proposition 2.7. Then

$$\langle g^*, \varphi\rangle = \fint_\Gamma \varphi \, d\mu_\Gamma \quad \text{for all} \quad \varphi \in \mathcal{L}^\infty(U)\,,$$

core $\mu_\Gamma \subset \Gamma$ by (3.12), and $|\mu_\Gamma|(U) = \|g^*\|$.

For $f \in \mathcal{L}^\infty(U)$ we now consider $f_\Gamma^* \in \mathcal{L}^\infty(U)^*$ related to $f\mu_\Gamma$ and we have

$$\langle f_\Gamma^*, \varphi\rangle = \langle g^*, \varphi f\rangle = \fint_\Gamma \varphi f \, d\mu_\Gamma \quad \text{for all} \quad \varphi \in \mathcal{L}^\infty(U)$$

(cf. (2.6)). Obviously f_Γ^* is also a trace on Γ. From (3.11) we obtain (3.5). Clearly, the mapping T is linear and, by $\|Tf\| \le \|f\|_{\mathcal{L}^\infty}|\mu_\Gamma|(U)$, also continuous.

For the last statement we fix $f, \varphi \in \mathcal{L}^\infty(U)$ and set

$$\beta^i := \liminf_{\delta\downarrow 0} \frac{1}{\gamma(\delta)} \int_{\Gamma_\delta \cap E} \varphi f \, d\mathcal{L}^n\,, \quad \beta^s := \limsup_{\delta\downarrow 0} \frac{1}{\gamma(\delta)} \int_{\Gamma_\delta \cap E} \varphi f \, d\mathcal{L}^n\,.$$

By (3.5) we have

$$\beta^i \le \beta := \fint_\Gamma \varphi f \, d\mu_\Gamma \le \beta^s\,.$$

If $\beta = \beta^i$ or $\beta = \beta^s$ we use the definition of $\liminf$ or $\limsup$, respectively, to get the assertion. For $\beta \in (\beta^i, \beta^s)$ we first observe that

$$\delta \to I(\delta) := \frac{1}{\gamma(\delta)} \int_{\Gamma_\delta \cap E} \varphi f \, d\mathcal{L}^n$$

is continuous for $\delta > 0$. Hence the mapping I attains the value γ on each interval $(0, \tilde\delta)$ with $\tilde\delta > 0$. But this implies the statement. □

Proof of Corollary 3.2 Let us fix $\varepsilon > 0$. By (3.8) there is some k_0 such that for all $k > k_0$ and all $\delta \in (0, \tilde\delta)$

$$\frac{1}{\gamma(\delta)} \int_{\Gamma_\delta \cap E} \varphi \, d\mathcal{L}^n - \varepsilon \le \frac{1}{\gamma(\delta)} \int_{\Gamma_\delta \cap E} \varphi_k \, d\mathcal{L}^n \le \frac{1}{\gamma(\delta)} \int_{\Gamma_\delta \cap E} \varphi \, d\mathcal{L}^n + \varepsilon\,.$$

Using the limit from (3.6) with $f \equiv 1$ and using (3.9) we obtain

$$\fint_\Gamma \varphi \, d\mu_\Gamma - \varepsilon \le \fint_\Gamma \varphi_k \, d\mu_\Gamma \le \fint_\Gamma \varphi \, d\mu_\Gamma + \varepsilon \quad \text{for all } k > k_0\,.$$

The arbitrariness of $\varepsilon > 0$ implies the assertion. □

3.2 Traces of Vector Fields with Divergence Measure

In our further treatment we are interested in traces that describe the behavior near the boundary $\partial\Omega$ for vector fields where the distributional divergence is a Radon measure. As special cases we consider Sobolev functions and BV functions. In our subsequent treatment we always assume that $U \subset \mathbb{R}^n$ is an open set.

Let us first recall some notation. For vector fields $F = (F_1, \dots, F_m)$ we use

$$\mathcal{L}^p(U, \mathbb{R}^m) := \mathcal{L}^p(U)^m, \quad \mathcal{W}^{1,p}(U, \mathbb{R}^m) := \mathcal{W}^{1,p}(U)^m$$

with the norms

$$\begin{aligned} \|F\|_p = \|F\|_{\mathcal{L}^p} &:= \Big(\int_U |F|^p \, d\mathcal{L}^n \Big)^{\frac{1}{p}} \quad \text{for} \quad 1 \le p < \infty, \\ \|F\|_{\mathcal{W}^{1,p}} &:= \big(\|F\|_p^p + \|DF\|_p^p \big)^{\frac{1}{p}} \quad \text{for } 1 \le p < \infty, \\ \|F\|_\infty = \|F\|_{\mathcal{L}^\infty} &:= \operatorname*{ess\,sup}_U |F|, \\ \|F\|_{\mathcal{W}^{1,\infty}} &:= \max\big\{ \|F\|_\infty, \|DF\|_\infty \big\} \end{aligned} \tag{3.13}$$

(where $|\cdot|$ is the Euclidean norm and DF is interpreted as mn-vector). Moreover

$$\begin{aligned} \|f\|_{\mathcal{BV}} &:= \|f\|_1 + |Df|(U) \quad \text{for} \quad f \in \mathcal{BV}(U), \\ \|\mu\| &:= |\mu|(U) \quad \text{for} \quad \mu \in \mathrm{ba}\big(U, \mathcal{B}(U), \mathcal{L}^n\big)^m . \end{aligned}$$

Proposition 2.7 tells us that $\mathrm{ba}\big(U, \mathcal{B}(U), \mathcal{L}^n\big)^m$ is the dual of $\mathcal{L}^\infty(U, \mathbb{R}^m)$.

We say that $F \in \mathcal{L}^1_{\mathrm{loc}}(U, \mathbb{R}^n)$ has *divergence measure* if there is a (signed) Radon measure on U denoted by $\operatorname{div} F$ such that

$$\int_U F \cdot D\varphi \, d\mathcal{L}^n = - \int_U \varphi \, d \operatorname{div} F \quad \text{for all} \quad \varphi \in C_c^\infty(U) \tag{3.14}$$

(i.e. the distributional divergence of F is a signed Radon measure). By approximation, (3.14) is even valid for all $\varphi \in \mathcal{W}^{1,\infty}(U)$ having compact support in U (take mollifications $\varphi_n \in C_c^\infty(U)$ of φ such that $D\varphi_n \to D\varphi$ $\mathcal{L}^n$-a.e. on U and use dominated convergence on the left hand side). The space of vector fields in $\mathcal{L}^p$ having divergence measure is defined by

$$\mathcal{DM}^p(U) := \big\{ F \in \mathcal{L}^p(U, \mathbb{R}^n) \;\big|\; |\operatorname{div} F|(U) < \infty \big\}, \quad 1 \le p \le \infty,$$

where the total variation $|\operatorname{div} F|(U)$ equals

$$|\operatorname{div} F|(U) = \sup\left\{ \int_U F \cdot D\varphi \, d\mathcal{L}^n \;\middle|\; \varphi \in C_c^1(U),\ |\varphi| \le 1 \right\}.$$

We have that $\mathcal{DM}^p(U)$ is a Banach space with the norm

$$\|F\|_{\mathcal{DM}^p} := \|F\|_{\mathcal{L}^p} + |\operatorname{div} F|(U).$$

For a Borel set $\Gamma \subset \overline{U}$ we use the semi norm on $\mathcal{L}^\infty(U, \mathbb{R}^m)$ given by

$$\|\varphi\|_\Gamma := \lim_{\delta \downarrow 0} \|\varphi\|_{\mathcal{L}^\infty(\Gamma_\delta \cap U, \mathbb{R}^m)}$$

(cf. (2.10)). With the subspace

$$Z := \{\varphi \in \mathcal{L}^\infty(U, \mathbb{R}^m) \mid \|\varphi\|_\Gamma = 0\} \tag{3.15}$$

we define the factor space

$$\mathcal{L}^\infty_\Gamma(U, \mathbb{R}^m) := \mathcal{L}^\infty(U, \mathbb{R}^m)/Z. \tag{3.16}$$

The equivalence class containing $\varphi \in \mathcal{L}^\infty(U, \mathbb{R}^m)$ is denoted by

$$\varphi_{\wr\Gamma} := \varphi + Z.$$

This way we can describe φ infinitesimally close to Γ, but not at Γ. If φ is continuous on a neighborhood of Γ with a continuous extension up to Γ, we can identify $\varphi_{\wr\Gamma}$ with the restriction $\varphi_{|\Gamma}$.

Now we are able to provide a large class of traces that will be the basis for upcoming general Gauss-Green formulas.

Theorem 3.3 *Let $U \subset \mathbb{R}^n$ be open and bounded and let $\Omega \in \mathcal{B}(U)$. Then there is a linear continuous operator $T : \mathcal{DM}^1(U) \to \mathcal{W}^{1,\infty}(U)^*$ such that*

$$\langle TF, \varphi \rangle = \operatorname{div}(\varphi F)(\Omega) = \int_\Omega \varphi \, d \operatorname{div} F + \int_\Omega D\varphi \cdot F \, d\mathcal{L}^n \tag{3.17}$$

for all $\varphi \in \mathcal{W}^{1,\infty}(U)$ and TF is a trace on $\partial\Omega$ over $\mathcal{W}^{1,\infty}(U)$ for all vector fields $F \in \mathcal{DM}^1(U)$. Moreover

$$\langle TF, \varphi \rangle = 0 \quad \textit{if} \quad \varphi_{|(\partial\Omega)_\delta \cap \Omega} = 0$$

for some $\delta > 0$.

We call T *trace operator*. In Theorem 4.1 below we will exploit the structure of $\mathcal{W}^{1,\infty}(U)^*$ to get a general representation for these traces. The proofs of the theorem and of the subsequent results are collected at the end of this section.

Remark 3.4 The functional $T^* \in \mathcal{DM}^1(U)^*$ given by

$$\langle T^*, F\rangle = \operatorname{div} F(\Omega)$$

is a trace on $\partial\Omega$ over $\mathcal{DM}^1(U)$ as one can see similar to the proof of the theorem. Thus we could take $\langle T^*, \varphi F\rangle$ instead of $\langle TF, \varphi\rangle$ in (3.17). The advantage would be to have merely one functional T^* for all F. However, the lack of knowledge about the structure of $\mathcal{DM}^1(U)^*$ prevents a direct representation of traces that way in general.

Corollary 3.5 *Let $U \subset \mathbb{R}^n$ be open and bounded, $\Omega \in \mathcal{B}(U)$, $F \in \mathcal{DM}^1(U)$, and let T be as in* (3.17). *If $\varphi \in \mathcal{W}^{1,\infty}(U)$ with*

$$\varphi_{\wr\partial\Omega} = 0 \qquad (\textit{i.e. } \|\varphi\|_{\partial\Omega} = 0)$$

and

$$(D\varphi)_{\wr\partial\Omega} = 0 \qquad \textit{or} \qquad \mathcal{L}^n(\Omega \setminus \operatorname{int}\Omega) = 0\,, \tag{3.18}$$

then $\langle TF, \varphi\rangle = 0$.

Notice that (3.18) is satisfied if Ω is open or if $\mathcal{L}^n(\partial\Omega) = 0$. For $U = \Omega$ open and bounded the previous result is similar to Theorem 2.3 in Šilhavý [43] where the right hand side in (3.17) is considered as functional over bounded $\varphi \in \operatorname{Lip}(\mathbb{R}^n)$ and it is shown that this functional agrees with a linear continuous functional on $\operatorname{Lip}(\partial\Omega)$.

Remark 3.6 Corollary 3.5 readily implies that the trace TF from Theorem 3.3 is uniquely determined if it is known for all $\varphi \in \mathcal{W}^{1,\infty}(U)$ with support in $(\partial\Omega)_\delta$ for some $\delta > 0$.

As simple consequence of Theorem 3.3 we get some analogous statement for Sobolev functions and BV functions.

Proposition 3.7 *Let $U \subset \mathbb{R}^n$ be open and bounded and let $\Omega \in \mathcal{B}(U)$. Then there is a linear continuous operator $T : \mathcal{BV}(U) \to \mathcal{W}^{1,\infty}(U, \mathbb{R}^n)^*$ such that*

$$\langle Tf, \varphi\rangle = \operatorname{div}(f\varphi)(\Omega) = \int_\Omega f \operatorname{div}\varphi \, d\mathcal{L}^n + \int_\Omega \varphi \, dDf \tag{3.19}$$

for all $\varphi \in \mathcal{W}^{1,\infty}(U, \mathbb{R}^n)$. Moreover, Tf is a trace on $\partial\Omega$ over $\mathcal{W}^{1,\infty}(U, \mathbb{R}^n)$ for all functions $f \in \mathcal{BV}(U)$. If

$$\varphi_{|(\partial\Omega)_\delta \cap \Omega} = 0 \textit{ for some } \delta > 0$$

or if

$$\varphi_{\wr\partial\Omega} = 0 \quad \textit{and (3.18) is satisfied,}$$

then we have $\langle Tf, \varphi\rangle = 0$.

Remark 3.8

(1) For function $f \in \mathcal{BV}(U)$ the distributional partial derivatives $D_{x_k} f$ are signed Radon measures and the distributional gradient Df is the vector-valued Radon measure

$$Df = (D_{x_1} f, \ldots, D_{x_n} f) \tag{3.20}$$

(cf. [2, p. 117]). Thus the most right integral in (3.19) has to be taken as

$$\int_\Omega \varphi \, dDf = \sum_{k=1}^{n} \int_\Omega \varphi^k \, dD_{x_k} f \quad \text{where} \quad \varphi = (\varphi^1, \ldots, \varphi^n).$$

(2) Proposition 3.7 is valid for all Sobolev functions $f \in \mathcal{W}^{1,1}(U)$, since they belong to $\mathcal{BV}(U)$ (cf. [22, p. 170]). For such f the measure Df equals $Df(\cdot)\mathcal{L}^n$ with the weak gradient $Df(\cdot)$ as density. Therefore

$$|Df|(\Omega) = \int_\Omega |Df| \, d\mathcal{L}^n, \quad \|f\|_{\mathcal{BV}} = \|f\|_{\mathcal{W}^{1,1}},$$

and in (3.19) we can replace

$$\int_\Omega \varphi \, dDf = \int_\Omega \varphi \cdot Df \, d\mathcal{L}^n.$$

(3) Let $U \subset \mathbb{R}^n$ be open, bounded and let $\Omega \in \mathcal{B}(U)$. We consider the space

$$X := \{f \in \mathcal{W}^{1,1}(U) \mid Df \in \mathcal{DM}^1(U)\}, \quad \|f\|_X := \|f\|_{\mathcal{L}^1} + \|Df\|_{\mathcal{DM}^1}.$$

This means that Δf in the sense of distributions is a Radon measure. Now we define $T : X \to \mathcal{W}^{1,\infty}(U, \mathbb{R}^n)^*$ by

$$\langle Tf, \varphi\rangle = \int_\Omega \varphi \, d\Delta f + \int_\Omega Df \cdot D\varphi \, d\mathcal{L}^n.$$

Theorem 3.3 implies that Tf is a trace on $\partial\Omega$ and we readily conclude that T is continuous. For Ω open with Lipschitz boundary and f smooth we obviously have

$$\langle Tf, \varphi \rangle = \int_{\partial\Omega} \varphi Df \cdot \nu^{\Omega} \, d\mathcal{H}^{n-1} .$$

Proof of Theorem 3.3 First we note that

$$\varphi F \in \mathcal{DM}^1(U) \quad \text{for all} \quad F \in \mathcal{DM}^1(U),\ \varphi \in \mathcal{W}^{1,\infty}(U) \tag{3.21}$$

and that, as measures,

$$\operatorname{div}(\varphi F) = \varphi \operatorname{div} F + F \cdot D\varphi \, \mathcal{L}^n \tag{3.22}$$

(cf. Proposition 2.2 and the subsequent comment in [43]). For $F \in \mathcal{DM}^1(U)$ we set

$$\langle TF, \varphi \rangle := \operatorname{div}(\varphi F)(\Omega) \quad \text{for all} \quad \varphi \in \mathcal{W}^{1,\infty}(U) .$$

From (3.22) we get (3.17) and

$$|\langle TF, \varphi \rangle| \le \|\varphi\|_\infty \, |\operatorname{div} F| \, (U) + \|D\varphi\|_\infty \|F\|_1 \le \|\varphi\|_{\mathcal{W}^{1,\infty}} \, \|F\|_{\mathcal{DM}^1} \, .$$

Hence $TF \in \mathcal{W}^{1,\infty}(U)^*$ with $\|TF\| \le \|F\|_{\mathcal{DM}^1}$. Therefore operator T is linear and continuous as stated.

We now consider $\varphi \in \mathcal{W}^{1,\infty}(U)$ with

$$\varphi_{|(\partial\Omega)_\delta \cap \Omega} = 0 \quad \text{for some} \quad \delta > 0 . \tag{3.23}$$

Then

$$D\varphi = 0 \quad \mathcal{L}^n\text{-a.e. on} \quad (\partial\Omega)_\delta \cap \Omega \tag{3.24}$$

(cf. [22, p. 130]). For $\delta' = \frac{\delta}{3}$ we define $\chi \in \mathcal{W}^{1,\infty}(\mathbb{R}^n)$ by

$$\chi(x) := \begin{cases} 1 & \text{for } x \in \Omega_{-2\delta'} , \\ 0 & \text{for } x \notin \Omega_{-\delta'} , \\ 1 - \frac{1}{\delta'} \operatorname{dist}_{(\Omega_{-2\delta'})}(x) & \text{otherwise} . \end{cases} \tag{3.25}$$

Clearly,

$$0 \le \chi \le 1 , \quad \chi = 1 \text{ on } \overline{\Omega_{-\delta}} , \quad \operatorname{supp} \chi \subset \overline{\Omega_{-\delta'}} \subset \overline{U_{-\delta'}} ,$$

$$(1 - \chi)\varphi = 0 \text{ on } \Omega , \quad \varphi D\chi = 0 \text{ on } U ,$$

$$(1 - \chi) D\varphi = 0 \quad \mathcal{L}^n\text{-a.e. on } \Omega .$$

Using (3.17) we obtain

$$\begin{aligned}\langle TF,\varphi\rangle &= \int_\Omega (1-\chi)\varphi\, d\operatorname{div} F + \int_\Omega (1-\chi) D\varphi\cdot F\, d\mathcal{L}^n \\ &\quad + \int_\Omega \chi\varphi\, d\operatorname{div} F + \int_\Omega \chi D\varphi\cdot F\, d\mathcal{L}^n \\ &= \int_U \chi\varphi\, d\operatorname{div} F + \int_U \big(\chi D\varphi + \varphi D\chi\big)\cdot F\, d\mathcal{L}^n \\ &= \int_U \chi\varphi\, d\operatorname{div} F + \int_U F\cdot D(\chi\varphi)\, d\mathcal{L}^n\,.\end{aligned}$$

Obviously $\chi\varphi \in \mathcal{W}^{1,\infty}(U)$ and it has compact support in U. The definition of divergence measure in (3.14) and the subsequent comment imply

$$\int_U \chi\varphi\, d\operatorname{div} F + \int_U F\cdot D(\chi\varphi)\, d\mathcal{L}^n = 0\,.$$

Consequently $\langle TF,\varphi\rangle = 0$ for all $\varphi \in \mathcal{W}^{1,\infty}(U)$ satisfying (3.23). This shows the last statement in the theorem and readily implies that TF is a trace on $\partial\Omega$ over $\mathcal{W}^{1,\infty}(U)$. □

Proof of Corollary 3.5 For $\delta > 0$ we use $\chi = \chi_\delta$ as in (3.25) and we have

$$0 \le \chi_\delta \le 1\,, \quad \chi_\delta = 1 \text{ on } \overline{\Omega_{-\delta}}\,, \quad \operatorname{supp}\chi_\delta \subset \overline{\Omega_{-\frac{\delta}{3}}} \subset \overline{U_{-\frac{\delta}{3}}}\,.$$

If $\varphi \in \mathcal{W}^{1,\infty}(U)$, then $\chi_\delta\varphi \in \mathcal{W}^{1,\infty}(U)$, it has compact support on U, and it vanishes outside Ω. Hence $D\chi_\delta = 0$ $\mathcal{L}^n$-a.e. on $U \setminus (\Omega \setminus \Omega_{-\delta})$ (cf. [22, p. 130]). Therefore the definition of divergence measure in (3.14) and the subsequent comment give for all $\delta > 0$

$$\begin{aligned} 0 &= \int_U \chi_\delta\varphi\, d\operatorname{div} F + \int_U F\cdot D(\chi_\delta\varphi)\, d\,\mathcal{L}^n \\ &= \int_\Omega \chi_\delta\varphi\, d\operatorname{div} F + \int_\Omega \chi_\delta F\cdot D\varphi\, d\,\mathcal{L}^n + \int_\Omega \varphi F\cdot D\chi_\delta\, d\,\mathcal{L}^n\,.\end{aligned}$$

Consequently

$$\begin{aligned}\langle TF,\varphi\rangle &= \int_\Omega (1-\chi_\delta)\varphi\, d\operatorname{div} F + \int_\Omega (1-\chi_\delta) D\varphi\cdot F\, d\,\mathcal{L}^n \\ &\quad + \int_\Omega \chi_\delta\varphi\, d\operatorname{div} F + \int_\Omega \chi_\delta D\varphi\cdot F\, d\,\mathcal{L}^n\end{aligned}$$

$$
\begin{aligned}
&= \int_\Omega (1-\chi_\delta)\varphi \, d \operatorname{div} F + \int_\Omega (1-\chi_\delta) D\varphi \cdot F \, d\mathcal{L}^n \\
&\quad - \int_\Omega \varphi F \cdot D\chi_\delta \, d\mathcal{L}^n . \qquad (3.26)
\end{aligned}
$$

Let now $\varphi_{\wr\partial\Omega} = 0$. Then $\varphi(x) = 0$ for $x \in \partial\Omega \cap \Omega$ by continuity of φ on U. For $x \in \operatorname{int} \Omega \setminus \Omega_{-\delta}$ there is $x' \in \partial\Omega$ such that

$$
|x - x'| = \operatorname{dist}_{\partial\Omega}(x) \le \delta
$$

and, consequently,

$$
B_{|x-x'|}(x) \subset \Omega .
$$

For any $\delta' > 0$ we can find $x'' \in (x', x)$ (open segment connecting x, x') such that $|\varphi(x'')| < \delta'$ by $\|\varphi\|_{\partial\Omega} = 0$. Therefore

$$
|\varphi(x)| \le |\varphi(x) - \varphi(x'')| + |\varphi(x'')| \le \delta \|D\varphi\|_\infty + \delta' .
$$

The arbitrariness of δ' and x implies

$$
|\varphi(x)| \le \delta \|D\varphi\|_\infty \quad \text{on} \quad \Omega \setminus \Omega_{-\delta} .
$$

Using $|D\chi_\delta| \le \frac{3}{\delta}$ $\mathcal{L}^n$-a.e. on U and $D\chi_\delta = 0$ $\mathcal{L}^n$-a.e. on $U \setminus \big(\Omega \setminus \Omega_{-\delta}\big)$ we get

$$
|\varphi D\chi_\delta| \le 3 \|D\varphi\|_\infty \quad \mathcal{L}^n\text{-a.e. on} \quad U .
$$

Since $D\chi_\delta(x) \to 0$ for all $x \in U$, dominated convergence gives

$$
\int_U \varphi F \cdot D\chi_\delta \, d\mathcal{L}^n \to 0 \quad \text{as} \quad \delta \to 0 .
$$

From $\varphi_{\wr\partial\Omega} = 0$ we also obtain that

$$
\big|(1 - \chi_\delta(x))\varphi(x)\big| \le \|\varphi_{|\Omega\setminus\Omega_{-\delta}}\|_\infty \overset{\delta\to 0}{\longrightarrow} 0 \quad \text{for all} \quad x \in \Omega
$$

and, therefore,

$$
\int_\Omega (1 - \chi_\delta)\varphi \, d \operatorname{div} F \to 0 \quad \text{as} \quad \delta \to 0 .
$$

Analogously, $(D\varphi)_{\wr\partial\Omega} = 0$ implies $\big|(1-\chi_\delta(x))D\varphi(x)\big| \overset{\delta\to 0}{\longrightarrow} 0$ for all $x \in \Omega$. Thus

$$\int_\Omega (1-\chi_\delta)F \cdot D\varphi \, d\mathcal{L}^n \to 0 \quad \text{as} \quad \delta \to 0\,.$$

The last convergence is also obtained in the case where $\mathcal{L}^n(\Omega \setminus \operatorname{int}\Omega) = 0$, since then $\chi_\delta \to 1$ $\mathcal{L}^n$-a.e. on Ω.

Now we can take the limit $\delta \to 0$ in (3.26) to get $\langle TF, \varphi\rangle = 0$. □

Proof of Proposition 3.7 For $f \in \mathcal{BV}(U)$ we consider the vector fields

$$F_k = (F_k^1, \dots, F_k^n) \quad \text{with} \quad F_k^k = f, \;\; F_k^j = 0 \;\; \text{for} \;\; k \neq j$$

where $k = 1, \dots, n$. Obviously $F_k \in \mathcal{DM}^1(U)$ for all k and, with the notation from (3.20),

$$\begin{aligned} \|F_k\|_{\mathcal{DM}^1} &= \|F_k\|_{\mathcal{L}^1} + |\operatorname{div} F_k|(U) \\ &= \|f\|_{\mathcal{L}^1} + |D_{x_k} f|(U) \\ &\leq \|f\|_{\mathcal{L}^1} + |Df|(U) = \|f\|_{\mathcal{BV}}\,. \end{aligned}$$

Hence, by Theorem 3.3, there are linear and continuous mappings

$$T_k : \mathcal{BV}(U) \to \mathcal{W}^{1,\infty}(U, \mathbb{R}^n)^* \quad (k = 1, \dots, n)$$

such that each $T_k f$ is a trace on $\partial\Omega$ and such that

$$\langle T_k f, \varphi\rangle = \operatorname{div}(\varphi^k F_k)(\Omega) = \int_\Omega \varphi^k \, dD_{x_k} f + \int_\Omega f \varphi^k_{x_k} \, d\mathcal{L}^n$$

for all $\varphi = (\varphi^1, \dots, \varphi^n) \in \mathcal{W}^{1,\infty}(U, \mathbb{R}^n)$. Summing over k we get the first statement of the proposition for $T = \sum_k T_k$. The assertions related to $\langle Tf, \varphi\rangle = 0$ follow directly from Theorem 3.3 and Corollary 3.5. □

3.3 Representation of Traces

For a powerful theory we now need suitable representations for the traces introduced in the previous section. In general we are interested in traces over $\mathcal{W}^{1,\infty}(U, \mathbb{R}^m)$ which requires representations of functionals in $\mathcal{W}^{1,\infty}(U, \mathbb{R}^m)^*$. To avoid interruptions in the presentation of the subsequent essential results, we collect all proofs in the next Sect. 3.4.

Let us start with some preliminary considerations. First we consider the semi norm $\|\varphi\|_\Gamma$ and the related factor space $\mathcal{L}^\infty_\Gamma(U,\mathbb{R}^m)$ introduced in (3.16).

Lemma 3.9 *Let $U \subset \mathbb{R}^n$ be open and let $\Gamma \subset \overline{U}$ be a Borel set. Then the subspace*

$$Z = \{\varphi \in \mathcal{L}^\infty(U,\mathbb{R}^m) \mid \|\varphi\|_\Gamma = 0\}$$

(cf. (3.15)) is closed and, for all $\varphi \in \mathcal{L}^\infty(U,\mathbb{R}^m)$,

$$\|\varphi\|_\Gamma = \operatorname{dist}_Z \varphi := \inf_{\psi \in Z} \|\varphi - \psi\|_{\mathcal{L}^\infty(U)}, \tag{3.27}$$

$$\|\varphi\|_\Gamma = \|\varphi + \psi\|_\Gamma \quad \text{if} \quad \psi \in Z. \tag{3.28}$$

Moreover

$$\mathcal{L}^\infty_\Gamma(U,\mathbb{R}^m) = \big\{\varphi_{\wr\Gamma} \;\big|\; \varphi_{\wr\Gamma} = \varphi + Z,\ \varphi \in \mathcal{L}^\infty(U,\mathbb{R}^m)\big\}$$

is a Banach space with the norm $\|\varphi_{\wr\Gamma}\| := \|\varphi\|_\Gamma$ and

$$\mathcal{L}^\infty_\Gamma(U,\mathbb{R}^m)^* \cong \{f^* \in \mathcal{L}^\infty(U,\mathbb{R}^m)^* \mid \langle f^*,\varphi\rangle = 0 \text{ for all } \varphi \in Z\}$$

as isometric isomorphism. For each $f^*_\Gamma \in \mathcal{L}^\infty_\Gamma(U,\mathbb{R}^m)^*$ there is some vector-valued measure $\mu \in \mathrm{ba}\big(U,\mathcal{B}(U),\mathcal{L}^n\big)^m$ such that

$$\langle f^*_\Gamma, \varphi_{\wr\Gamma}\rangle = \int_U \varphi\, d\mu \quad \text{for all } \varphi_{\wr\Gamma} \in \mathcal{L}^\infty_\Gamma(U,\mathbb{R}^m)$$

and

$$\int_U \varphi\, d\mu = 0 \quad \text{for all } \varphi \in Z \ \ (\text{i.e. } \|\varphi\|_\Gamma = 0).$$

If U is bounded, then $\operatorname{core}\mu \subset \overline{\Gamma}$.

For the characterization of $\mathcal{W}^{1,\infty}(U,\mathbb{R}^m)^*$ we identify $\mathcal{W}^{1,\infty}(U,\mathbb{R}^m)$ with a subspace of a product space of the form

$$\big\{(\varphi, D\varphi) \in X_0 \times \mathcal{L}^\infty(U,\mathbb{R}^{mn}) \;\big|\; \varphi \in \mathcal{W}^{1,\infty}(U,\mathbb{R}^m)\big\}$$

where X_0 is a suitable Banach space (e.g. $\mathcal{L}^\infty(U,\mathbb{R}^m)$ in the general case or $C(\Gamma,\mathbb{R}^m)$ if all φ are continuously extendable up to $\Gamma \subset \overline{U}$).

Lemma 3.10 *Let X_0 be a Banach space with $\|.\|_{X_0}$ and let $U \subset \mathbb{R}^n$ be open. Then*

$$X = X_0 \times \mathcal{L}^\infty(U,\mathbb{R}^{mn}) \quad \text{with} \quad \|(\varphi,\Phi)\|_X = \max\big\{\|\varphi\|_{X_0}, \|\Phi\|_{\mathcal{L}^\infty}\big\}$$

is a Banach space and X^* is isometrically isomorphic to

$$X_0^* \times \mathrm{ba}\big(U, \mathcal{B}(U), \mathcal{L}^n\big)^{mn} \quad \text{with} \quad \|(f^*, \mu)\| = \|f^*\| + \|\mu\|$$

such that

$$\langle (f^*, \mu), (\varphi, \Phi) \rangle = \langle f^*, \varphi \rangle + \int_U \Phi \, d\mu \tag{3.29}$$

for all $(\varphi, \Phi) \in X$, $(f^*, \mu) \in X^*$ and $\|\mu\| = |\mu|(U)$.

It turns out that the lemma can be used for the characterization of $\mathcal{W}^{1,\infty}(U, \mathbb{R}^m)^*$, if there is a Banach space X_0 such that $\mathcal{W}^{1,\infty}(U, \mathbb{R}^m)$ can be identified with some subspace of X.

Proposition 3.11 *Let X_0 be a Banach space, let $U \subset \mathbb{R}^n$ be open, and assume that there is a linear mapping $\iota_0 : \mathcal{W}^{1,\infty}(U, \mathbb{R}^m) \to X_0$ such that*

$$\iota : \mathcal{W}^{1,\infty}(U, \mathbb{R}^m) \to X_0 \times \mathcal{L}^\infty(U, \mathbb{R}^{mn}) \quad \text{with} \quad \iota(\varphi) = (\iota_0(\varphi), D\varphi)$$

has a continuous inverse ι^{-1} on its image $\iota\big(\mathcal{W}^{1,\infty}(U, \mathbb{R}^m)\big)$ equipped with

$$\|\iota(\varphi)\| = \max\left\{\|\iota_0(\varphi)\|_{X_0}, \|D\varphi\|_\infty\right\}.$$

Then for each $f^* \in \mathcal{W}^{1,\infty}(U, \mathbb{R}^m)^*$ there are

$$f_0^* \in X_0^* \quad \text{and} \quad \mu \in \mathrm{ba}\big(U, \mathcal{B}(U), \mathcal{L}^n\big)^{mn}$$

such that

$$\langle f^*, \varphi \rangle = \langle f_0^*, \iota_0(\varphi) \rangle + \int_U D\varphi \, d\mu$$

for all $\varphi \in \mathcal{W}^{1,\infty}(U, \mathbb{R}^m)$. Moreover

$$\begin{aligned} \|(f_0^*, \mu)\| = \|f_0^*\| + \|\mu\| &= \|f^* \circ \iota^{-1}\| \\ \le \|\iota^{-1}\| \|f^*\| &= \|\iota^{-1}\| \sup_{\substack{\varphi \in \mathcal{W}^{1,\infty}(U, \mathbb{R}^m) \\ \|\varphi\|_{\mathcal{W}^{1,\infty}} \le 1}} \langle f^*, \varphi \rangle . \end{aligned} \tag{3.30}$$

Remark 3.12 The assumption for ι is satisfied if $\iota\big(\mathcal{W}^{1,\infty}(U, \mathbb{R}^m)\big)$ is closed and ι is continuous and injective, since then ι^{-1} is continuous by the open mapping

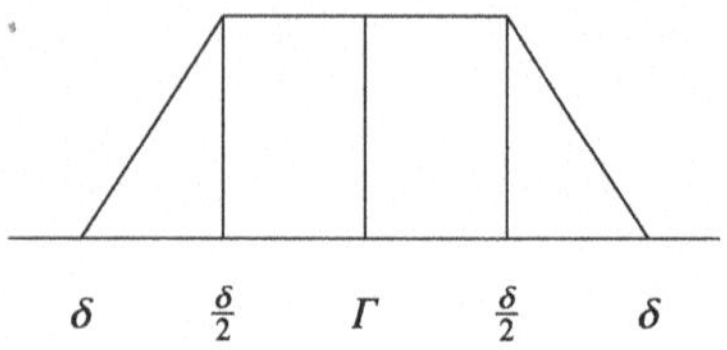

Fig. 3.1 Tent function χ_δ^Γ of Γ and δ

theorem. Alternatively an estimate

$$\|\varphi\|_{\mathcal{W}^{1,\infty}} \leq \tilde{c}\, \|\iota(\varphi)\| \quad \text{for all } \ \varphi \in \mathcal{W}^{1,\infty}(U, \mathbb{R}^m)$$

with some $\tilde{c} \geq 0$ is equivalent to existence and continuity of ι^{-1}.

For $X_0 = \mathcal{L}^\infty(U, \mathbb{R}^m)$ and $\iota_0(\varphi) = \varphi$, the assumptions of Proposition 3.11 are satisfied and we obtain a representation of $\mathcal{W}^{1,\infty}(U, \mathbb{R}^m)^*$ related to measures on U without further specification of their core. At this point we have to realize that the representation of $f^* \in \mathcal{W}^{1,\infty}(U, \mathbb{R}^m)^*$ by means of measures is not unique in general and that even traces on $\Gamma \subset \overline{U}$ over $\mathcal{W}^{1,\infty}(U, \mathbb{R}^m)$ can be represented with measures that are supported on all of U. Indeed, if we take the trace $f^* = TF$ from (3.17), then the right hand side of (3.17) itself gives a representation of it related to the measures

$$\operatorname{div} F \in \mathcal{M}(U) \quad \text{and} \quad F\mathcal{L}^n \in \operatorname{ba}\big(U, \mathcal{B}(U), \mathcal{L}^n\big)^n .$$

These measures might be provided by Proposition 3.11 for f^* if we consider $\operatorname{div} F$ as extension to some element of $\operatorname{ba}\big(U, \mathcal{B}(U), \mathcal{L}^n\big)$ according to Proposition 2.13. However, having in mind usual Gauss-Green formulas, we are rather interested in representations of traces on Γ with measures having core on or at least near Γ.

For such a localization we use the *tent function* $\chi_\delta^\Gamma : \mathbb{R}^n \to \mathbb{R}$ of Γ and $\delta > 0$ given by

$$\chi_\delta^\Gamma := \chi_{\Gamma_{\delta/2}} + \chi_{\Gamma_\delta \setminus \Gamma_{\delta/2}} \big(2 - \tfrac{2}{\delta} \operatorname{dist}_\Gamma\big) \tag{3.31}$$

which is 1 on $\Gamma_{\frac{\delta}{2}}$ and 0 outside Γ_δ (cf. Fig. 3.1). Note that $\chi_\delta^\Gamma \in \mathcal{W}^{1,\infty}(U)$, since it is Lipschitz continuous on $\mathbb{R}^n$. Let us mention that other choices of χ_δ^Γ provide the same results as long as χ_δ^Γ is Lipschitz continuous with support on $\overline{\Gamma_\delta}$ and with $\chi_\delta^\Gamma = 1$ on some neighborhood of Γ.

Proposition 3.13 *Let $U \subset \mathbb{R}^n$ be open, let $\Gamma \subset \overline{U}$ be closed, let $\delta > 0$, and let $f^* \in \mathcal{W}^{1,\infty}(U, \mathbb{R}^m)^*$ be a trace on Γ. Then there is some*

$$f_\delta^* \in \mathcal{W}^{1,\infty}(\Gamma_\delta \cap U, \mathbb{R}^m)^* \ \text{ with } \ \langle f^*, \varphi\rangle = \langle f^*, \chi_\delta^\Gamma \varphi\rangle = \langle f_\delta^*, \varphi_{|\Gamma_\delta \cap U}\rangle \tag{3.32}$$

for all $\varphi \in \mathcal{W}^{1,\infty}(U, \mathbb{R}^m)$ *and*

$$\| f_\delta^* \| \le 2 \| \chi_\delta^\Gamma \|_{\mathcal{W}^{1,\infty}(U,\mathbb{R}^m)} \| f^* \| .$$

Now we are able to represent any trace f^* on Γ over $\mathcal{W}^{1,\infty}(U, \mathbb{R}^m)$ by a functional $f_\delta^* \in \mathcal{W}^{1,\infty}(\Gamma_\delta \cap U, \mathbb{R}^m)^*$. This certainly leads to sharper results for the core of the related measures. But notice that f_δ^* really can depend on $\delta > 0$ in general (cf. Example 4.10 below). However if the f_δ^* are bounded in some sense, then f^* can be represented by measures with core in Γ (cf. Proposition 3.15 below).

For a more precise analysis we apply Proposition 3.11 to f_δ^* of Proposition 3.13. Here three choices of X_0 with corresponding $\iota_0 : \mathcal{W}^{1,\infty}(U, \mathbb{R}^m) \to X_0$ turn out to be of particular relevance. We say that we have case (G), (L), or (C) for $\Gamma \subset \overline{U}$ and $\delta > 0$ if the assumption in Proposition 3.11 is satisfied with $\Gamma_\delta \cap U$ instead of U and with

(G) $X_0 = \mathcal{L}^\infty(\Gamma_\delta \cap U, \mathbb{R}^m)$ and $\iota_0(\varphi) = \varphi$,
(L) $X_0 = \mathcal{L}^\infty_\Gamma(\Gamma_\delta \cap U, \mathbb{R}^m)$ and $\iota_0(\varphi) = \varphi_{\wr\Gamma}$,
(C) $X_0 = C(\Gamma, \mathbb{R}^m)$ and $\iota_0(\varphi) = \varphi_{|\Gamma}$ where all $\varphi \in \mathcal{W}^{1,\infty}(\Gamma_\delta \cap U, \mathbb{R}^m)$ are assumed to be continuously extendable up to Γ.

We always have the *general case* (G), since here ι is an isometry for all $\delta > 0$. The other cases, that we call *Lebesgue case* (L) and *continuity case* (C), turn out to be a condition for the geometry of $\Gamma \subset \overline{U}$ related to δ. With Lemma 3.9 we readily get for bounded U that $f_0^* \in X_0^*$ from Proposition 3.11 corresponds, among others, to a measure λ where

$$\text{(G): } \operatorname{core}\lambda \subset \overline{\Gamma_\delta \cap U}\,, \qquad \text{(L), (C): } \operatorname{core}\lambda \subset \overline{\Gamma}$$

and where λ can even be taken as σ-measure on Γ in the strongest case (C). This shows us the relevance of the different cases. Using these cases we now provide general representations of traces on Γ over $\mathcal{W}^{1,\infty}(U, \mathbb{R}^m)$. In Sect. 4.1 we combine these results with Theorem 3.3 to derive Gauss-Green formulas on arbitrary Borel sets $\Omega \subset U$ for vector fields in $\mathcal{DM}^1(U)$.

Theorem 3.14 *Let* $U \subset \mathbb{R}^n$ *be open and bounded, let* $\Gamma \subset \overline{U}$ *be closed, and let* $f^* \in \mathcal{W}^{1,\infty}(U, \mathbb{R}^m)^*$ *be a trace on* Γ. *Then for each* $\delta > 0$ *there are measures*

$$\lambda \subset \mathrm{ba}(U, \mathcal{B}(U), \mathcal{L}^n)^m \quad \textit{and} \quad \mu \in \mathrm{ba}(U, \mathcal{B}(U), \mathcal{L}^n)^{mn} \tag{3.33}$$

with

$$\operatorname{core}\lambda,\ \operatorname{core}\mu \subset \overline{\Gamma_\delta \cap U}$$

such that

$$\langle f^*, \varphi \rangle = \int_{\Gamma_\delta \cap U} \varphi \, d\lambda + \int_{\Gamma_\delta \cap U} D\varphi \, d\mu \tag{3.34}$$

for all $\varphi \in \mathcal{W}^{1,\infty}(U, \mathbb{R}^m)$ *and*

$$\begin{aligned} \|(\lambda, \mu)\| &= \|\lambda\| + \|\mu\| = |\lambda|(\Gamma_\delta \cap U) + |\mu|(\Gamma_\delta \cap U) \\ &\le c \sup_{\substack{\varphi \in \mathcal{W}^{1,\infty}(U,\mathbb{R}^m) \\ \|\varphi_{|\Gamma_\delta \cap U}\|_{\mathcal{W}^{1,\infty}} \le 1}} \langle f^*, \chi_\delta^\Gamma \varphi \rangle \end{aligned} \tag{3.35}$$

for a constant $c \ge 1$. *In the particular cases we have in addition that:*

(G): λ, μ *can be chosen such that equality holds in norm estimate* (3.35) *with* $c = 1$,
(L): λ, μ *can be chosen such that* $\operatorname{core} \lambda \subset \Gamma$ *and, thus,* λ *is independent of* δ *and* (3.34) *can be written as*

$$\langle f^*, \varphi \rangle = \fint_{\Gamma} \varphi \, d\lambda + \int_{\Gamma_\delta \cap U} D\varphi \, d\mu \,,$$

(C): λ, μ *can be chosen such that measure* λ *is related to a Radon measure* σ *with* $\operatorname{supp} \sigma \subset \Gamma$ *and, thus,* λ *is independent of* δ *and* (3.34) *becomes*

$$\langle f^*, \varphi \rangle = \int_{\Gamma} \varphi \, d\sigma + \int_{\Gamma_\delta \cap U} D\varphi \, d\mu$$

(where we always identify φ *with its continuous extension on* Γ*).*

Let us mention that the choice of the measures λ and μ in the theorem is not unique in general. In case (G) related to δ, the theorem is obviously true for λ, μ satisfying the assertion for $\delta' < \delta$. Moreover, for $\varphi \in \mathcal{W}^{1,\infty}(U, \mathbb{R}^m)$ the pairs $(\varphi, D\varphi)$ merely belong to a subspace of $\mathcal{L}^\infty(U)^m \times \mathcal{L}^\infty(U)^{nm}$ and there can be nontrivial measures $(\tilde{\lambda}, \tilde{\mu})$ that vanish on this subspace and that can always be added. Finally, the examples in Sect. 4.1 show how the contribution of λ and μ to the right hand side can be changed in certain situations.

Notice that the constant $c \ge 1$ in (3.35) depends on the particular choice of λ and μ. Though we always have $c = 1$ for some choice of λ, μ in case (G), we may only have (3.35) with a larger c for the choice of λ, μ in the other cases. More precisely, in cases (G), (L) and (C) we get from the proof that

$$c = \|\iota^{-1}\| \tag{3.36}$$

in (3.35) for the related ι according to Proposition 3.11. Before we discuss cases (L) and (C) in some more detail we consider important special cases of Theorem 3.14.

In general the measures $(\lambda, \mu) = (\lambda_\delta, \mu_\delta)$ on $\Gamma_\delta \cap U$ related to f^* depend on δ. But, if they are somehow bounded with respect to δ, they have a weak* cluster point giving a representation of f^* independent of δ.

Proposition 3.15 *Let $U \subset \mathbb{R}^n$ be open and bounded, let $\Gamma \subset \overline{U}$ be closed, and let $f^* \in \mathcal{W}^{1,\infty}(U, \mathbb{R}^m)^*$ be a trace on Γ. Then we can choose λ, μ in* (3.34) *such that* $\operatorname{core}\lambda, \operatorname{core}\mu \subset \Gamma$, *i.e. λ, μ are independent of δ and* (3.34) *becomes*

$$\langle f^*, \varphi\rangle = \fint_\Gamma \varphi\, d\lambda + \fint_\Gamma D\varphi\, d\mu,$$

if and only if

$$\liminf_{\delta\downarrow 0} \sup_{\substack{\varphi\in\mathcal{W}^{1,\infty}(U,\mathbb{R}^m)\\ \|\varphi_{|\Gamma_\delta\cap U}\|_{\mathcal{W}^{1,\infty}}\le 1}} \langle f^*, \chi_\delta^\Gamma \varphi\rangle < \infty\,. \tag{3.37}$$

If we have in addition case (C) *for some $\delta > 0$, then λ corresponds to a Radon measure supported on Γ.*

We call the trace f^* *finite* if (3.37) is satisfied. Notice that this condition gives some uniform bound for (λ, μ) with respect to δ according to (3.35). In the light of usual Gauss-Green formulas as in (3.1) it is also desirable to characterize the case where the measure μ can disappear in (3.34).

Proposition 3.16 *Let $U \subset \mathbb{R}^n$ be open and bounded, let $\Gamma \subset \overline{U}$ be closed, let $f^* \in \mathcal{W}^{1,\infty}(U, \mathbb{R}^m)^*$ be a trace on Γ.*

(1) *For $\delta > 0$ we can choose λ, μ in* (3.34) *such that $\mu = 0$, i.e.*

$$\langle f^*, \varphi\rangle = \int_{\Gamma_\delta\cap U} \varphi\, d\lambda \quad \textit{where} \quad \operatorname{core}\lambda \subset \overline{\Gamma_\delta \cap U}\,,$$

if and only if

$$\sup_{\substack{\varphi\in\mathcal{W}^{1,\infty}(U,\mathbb{R}^m)\\ \|\varphi_{|\Gamma_\delta\cap U}\|_{\mathcal{L}^\infty}\le 1}} \langle f^*, \varphi\rangle < \infty \tag{3.38}$$

is satisfied.

(2) *We can choose λ, μ in* (3.34) *such that $\mu = 0$ and* $\operatorname{core}\lambda \subset \Gamma$, *i.e.*

$$\langle f^*, \varphi\rangle = \fint_\Gamma \varphi\, d\lambda\,,$$

if and only if

$$\liminf_{\delta\downarrow 0} \sup_{\substack{\varphi\in\mathcal{W}^{1,\infty}(U,\mathbb{R}^m)\\ \|\varphi_{|\Gamma_\delta\cap U}\|_{\mathcal{L}^\infty}\le 1}} \langle f^*, \varphi\rangle < \infty\,. \tag{3.39}$$

If we have in addition case (C)*, then measure* λ *corresponds to a Radon measure that is supported on* Γ*.*

Condition (3.38) somehow says that the trace f^* can be considered as a linear continuous functional on $\mathcal{L}^\infty(\Gamma_\delta \cap U, \mathbb{R}^m)$. But notice that the possible choice $\mu = 0$ does not exclude other representations of f^* with $\mu \neq 0$ (cf. Example 4.14 below). The proof of (2) shows that (3.39) implies (3.37).

Let us come back to cases (L) and (C). Since they play an important role, we will provide some conditions that help to identify them. We say that $\Gamma_\delta \cap U$ is *bounded path connected with* Γ if there is a maximal length $\ell > 0$ such that for any $x \in \Gamma_\delta \cap U$ and any $\delta' > 0$ there are a point $y \in \Gamma_{\delta'} \cap U$ and a curve connecting x, y inside $\Gamma_\delta \cap U$ with length less than ℓ.

Proposition 3.17 *Let* $U \subset \mathbb{R}^n$ *be open, let* $\Gamma \subset \overline{U}$ *be closed, and let* $\delta > 0$*.*

(1) *If we have case* (L) *or* (C) *for* δ *and if* V *is a component of* $\Gamma_\delta \cap U$*, then*

$$\Gamma_{\delta'} \cap V \neq \emptyset \quad \textit{for all} \quad \delta' > 0\,.$$

(2) *If* $\Gamma_\delta \cap U$ *is bounded path connected with* Γ*, then we have case* (L) *for* δ *with* $c = 1 + \ell$ *in* (3.35).
(3) *If any* $\varphi \in \mathcal{W}^{1,\infty}(U, \mathbb{R}^m)$ *is continuously extendable up to* Γ *and if* $\Gamma_\delta \cap U$ *is bounded path connected with* Γ*, then we have case* (C) *for* δ *with* $c = 1 + \ell$ *in* (3.35).

These general assertions imply some important special cases.

Corollary 3.18 *Let* $U \subset \mathbb{R}^n$ *be open and bounded and let* $\Gamma \subset \overline{U}$ *be closed.*

(1) *If* $\Gamma = \partial U$*, then we have case* (L) *for any* $\delta > 0$ *and* $c \le 1 + \delta$ *in* (3.35).
(2) *If* $\Gamma = \partial U$ *and* U *has Lipschitz boundary, then we have case* (C) *for all* $\delta > 0$ *and* $c \le 1 + \delta$ *in* (3.35).
(3) *If* $\Gamma_\delta \Subset U$ *for* $\delta > 0$*, then we have case* (C) *for* δ *and* $c \le 1 + \delta$ *in* (3.35).

Let us elaborate on the cases (G), (L), (C) by applying Proposition 3.17 and Corollary 3.18 to some typical examples in $\mathbb{R}^2$.

Example 3.19 We consider $U, \Gamma \subset \mathbb{R}^2$ with

$$U := \big((0,1) \cup (1,2)\big) \times (0,1)\,, \qquad \Gamma := \{1\} \times (0,1)\,.$$

Since $\Gamma_\delta \cap U$ has two components, some $\varphi \in \mathcal{W}^{1,\infty}(\Gamma_\delta \cap U, \mathbb{R}^m)$ that equals different constants on each component cannot be extended continuously up to Γ. Thus we do not have (C) for any $\delta > 0$. But we readily verify the assumption of Proposition 3.17 (2) and, consequently, we have case (L) for all $\delta > 0$.

Example 3.20 In $\mathbb{R}^2$ we take

$$U := \bigcup_{k=1}^{\infty} R_k \quad \text{with} \quad R_k := \left(\tfrac{1}{2k+1}, \tfrac{1}{2k}\right) \times (0, 1), \qquad \Gamma := \{0\} \times (0, 1).$$

Obviously we cannot continuously extend all $\varphi \in \mathcal{W}^{1,\infty}(\Gamma_\delta \cap U, \mathbb{R}^m)$ up to Γ and thus, we do not have (C) for any $\delta > 0$.

For fixed $\delta > 0$ we now choose some $R_{k'} \subset \Gamma_\delta$. This is obviously a component of $\Gamma_\delta \cap U$ and clearly $\Gamma_{\delta'} \cap R_{k'} = \emptyset$ for all sufficiently small $\delta' > 0$. Hence we do not have (L) for any $\delta > 0$ by Proposition 3.17 (1). Therefore, for the treatment of a trace on Γ we can merely use the general case (G) in Theorem 3.14.

While U has infinite perimeter, we get the same results for some U with finite perimeter if we replace R_k and Γ with

$$\tilde{R}_k := \left(\tfrac{1}{2^{2k+1}}, \tfrac{1}{2^{2k}}\right) \times \left(0, \tfrac{1}{2^k}\right) \quad \text{and} \quad \tilde{\Gamma} := \{(0, 0)\}.$$

Example 3.21 With R_k as in the previous example we now choose in $\mathbb{R}^2$

$$U = B_2(0), \qquad \Omega = \bigcup_{k=1}^{\infty} R_k, \qquad \Gamma \subset \partial\Omega \text{ closed}.$$

In contrast to $\Gamma \subset \partial U$ in Example 3.20, we now have $\Gamma \Subset U$. This changes the situation essentially and, by Corollary 3.18 (3), we have case (C) for all small $\delta > 0$.

Example 3.22 In $\mathbb{R}^2$ we consider $\Gamma := \{0\} \times (0, 1)$ and

$$U := \big((0, 1) \times (0, 1)\big) \setminus \bigcup_{l=1}^{\infty} \Big(\left\{\tfrac{1}{2l}\right\} \times \left(0, \tfrac{3}{4}\right] \cup \left\{\tfrac{1}{2l+1}\right\} \times \left[\tfrac{1}{4}, 1\right)\Big)$$

(cf. Fig. 3.2). Obviously we cannot extend all $\varphi \in \mathcal{W}^{1,\infty}(\Gamma_\delta \cap U, \mathbb{R}^m)$ continuously up to Γ, since φ can oscillate between the inner boundaries. Therefore we do not have (C) for any $\delta > 0$. For treating (L) we first observe that $\Gamma_\delta \cap U$ is not bounded path connected with Γ for any $\delta > 0$. Thus we cannot use Proposition 3.17 and we have to check the assumption for ι in Proposition 3.11 directly. For that we fix

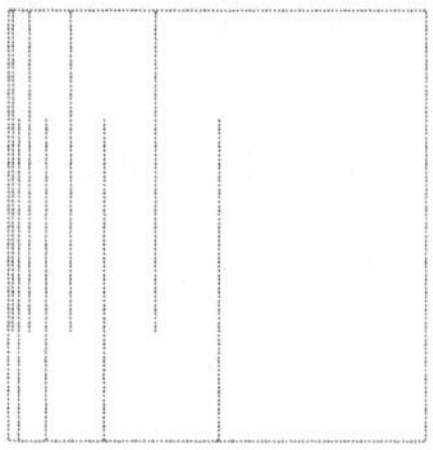

Fig. 3.2 The open set U

$\delta > 0$. Then for each $k \in \mathbb{N}$ there is some $\varphi_k \in \mathcal{W}^{1,\infty}(\Gamma_\delta \cap U)$ with

$$\|\varphi_k\|_\infty = k\,, \quad \|D\varphi_k\|_\infty = 1\,, \quad \varphi_{k \wr \Gamma} = 0$$

(roughly speaking, choose $\varphi_k = k$ near $\{\delta\} \times (0,1)$, decrease φ_k to zero towards Γ by respecting $\|D\varphi_k\|_\infty = 1$, and set $\varphi_k = 0$ in a small remaining neighborhood of Γ). Then

$$\|\varphi_k\|_{\mathcal{W}^{1,\infty}} = k+1\,, \quad \|\iota(\varphi_k)\| = \max\left\{\|\varphi_{k\wr\Gamma}\|,\ \|D\varphi_k\|_\infty\right\} = 1\,.$$

But this prevents continuity of ι^{-1} and hence we do not have (L) for any $\delta > 0$.

If we take $\Gamma' := \{1\} \times (0,1)$ instead of Γ, then $\Gamma'_\delta \cap U$ is bounded path connected with Γ' for $\delta < 1$ while it is not for $\delta \geq 1$. Thus we get (L) for $\delta < 1$ from Proposition 3.17 while we do not have (L) for $\delta \geq 1$ by arguments as above. For $\Gamma'' = \partial U$ we have (L) for all $\delta > 0$ by Corollary 3.18.

3.4 Proofs

Here we give the proofs of the results from the previous section.

Proof of Lemma 3.9 Since $\|\cdot\|_\Gamma$ is a semi norm, we get for $\varphi, \psi \in \mathcal{L}^\infty(U, \mathbb{R}^m)$

$$\big|\|\varphi\|_\Gamma - \|\psi\|_\Gamma\big| \leq \|\varphi - \psi\|_\Gamma \leq \|\varphi - \psi\|_{\mathcal{L}^\infty}\,. \tag{3.40}$$

Hence Z is a closed subspace. If $\psi \in Z$, then

$$\|\varphi\|_\Gamma \overset{(3.40)}{\leq} \|\varphi + \psi\|_\Gamma \leq \|\varphi\|_\Gamma + \|\psi\|_\Gamma = \|\varphi\|_\Gamma\,,$$

which implies (3.28). For $\varphi \in \mathcal{L}^\infty(U, \mathbb{R}^m)$ we define

$$\varphi_\delta := \begin{cases} \varphi \text{ on } U \setminus \Gamma_\delta\,, \\ 0 \text{ on } U \cap \Gamma_\delta \end{cases} \quad \text{for all } \delta > 0\,.$$

Obviously $\varphi_\delta \in Z$ for all δ and

$$\operatorname{dist}_Z \varphi \leq \inf_{\delta>0} \|\varphi - \varphi_\delta\|_{\mathcal{L}^\infty(U)} = \lim_{\delta \downarrow 0} \|\varphi\|_{\mathcal{L}^\infty(\Gamma_\delta \cap U)} = \|\varphi\|_\Gamma\,.$$

Assume that $\|\varphi\|_\Gamma > \operatorname{dist}_Z \varphi$, then there is $\psi \in Z$ with

$$\|\varphi - \psi\|_{\mathcal{L}^\infty} < \|\varphi\|_\Gamma \overset{(3.28)}{=} \|\varphi - \psi\|_\Gamma \leq \|\varphi - \psi\|_{\mathcal{L}^\infty}\,,$$

which is a contradiction. Hence $\|\varphi\|_\Gamma = \operatorname{dist}_Z \varphi$. Therefore, by standard results, $\mathcal{L}^\infty_\Gamma(U, \mathbb{R}^m)$ is a Banach space with $\|\varphi_{\wr\Gamma}\| = \|\varphi\|_\Gamma$ (cf. [55, p. 185]). By definition

$$Z^\perp := \{f^* \in \mathcal{L}^\infty(U, \mathbb{R}^m)^* \mid \langle f^*, \varphi\rangle = 0 \text{ for all } \varphi \in Z\}$$

is the annihilator of Z. For any $f^* \in Z^\perp$ we get an $f^*_\Gamma \in \mathcal{L}^\infty_\Gamma(U, \mathbb{R}^m)^*$ by

$$\langle f^*_\Gamma, \varphi_{\wr\Gamma}\rangle := \langle f^*, \varphi\rangle\,.$$

But this mapping defines an isometric isomorphism from $Z^\perp$ onto $\mathcal{L}^\infty_\Gamma(U, \mathbb{R}^m)^*$ (cf. [21, p. 72] or [49, p. 34, 99]).

Since $\mathcal{L}^\infty(U, \mathbb{R}^m)^*$ can be identified with $\operatorname{ba}\big(U, \mathcal{B}(U), \mathcal{L}^n\big)^m$, for the functional $f^*_\Gamma \in \mathcal{L}^\infty_\Gamma(U, \mathbb{R}^m)^*$ there is $\mu \in \operatorname{ba}\big(U, \mathcal{B}(U), \mathcal{L}^n\big)^m$ such that

$$\langle f^*_\Gamma, \varphi_{\wr\Gamma}\rangle = \int_U \varphi\, d\mu \quad \text{for all} \quad \varphi_{\wr\Gamma} \in \mathcal{L}^\infty_\Gamma(U, \mathbb{R}^m)$$

while

$$\int_U \varphi\, d\mu = 0 \quad \text{for all } \varphi \in Z\,. \tag{3.41}$$

Let now U be bounded and assume that $x \in \operatorname{core} \mu_k \setminus \overline{\Gamma}$ for some component μ_k of μ. Then there is some $\delta > 0$ and some open $V \subset \mathbb{R}^n \setminus \Gamma_{2\delta}$ containing x, such that $|\mu_k|(V \cap U) > 0$. Hence we can find some $\psi \in \mathcal{L}^\infty(U, \mathbb{R}^m)$ with $\psi_{|\Gamma_\delta} = 0$, with $\psi_j \equiv 0$ for $j \neq k$, and with $\int_U \psi\, d\mu > 0$. But this contradicts (3.41), since $\psi \in Z$. Therefore $\operatorname{core} \mu \subset \overline{\Gamma}$. □

Proof of Lemma 3.10 Obviously X is a Banach space. Moreover we have that $\operatorname{ba}\big(U, \mathcal{B}(U), \mathcal{L}^n\big)^{mn}$ with $\|\mu\| = |\mu|(U)$ is the dual of $\mathcal{L}^\infty(U, \mathbb{R}^{mn})$ by Proposition 2.7. Then X^* is the dual of X with (3.29) by standard arguments. For the norm of (f^*, μ) in X^* we fix $\varepsilon > 0$. Then there is $(\varphi^\varepsilon, \Phi^\varepsilon) \in X$ with

$$\|(\varphi^\varepsilon, \Phi^\varepsilon)\| \leq 1\,, \quad \|f^*\| \leq \langle f^*, \varphi^\varepsilon\rangle + \varepsilon\,, \quad \|\mu\| \leq \langle \mu, \Phi^\varepsilon\rangle + \varepsilon\,.$$

Hence, by (3.29),

$$\|(f^*, \mu)\| \leq \|f^*\| + \|\mu\| \leq \langle (f^*, \mu), (\varphi^\varepsilon, \Phi^\varepsilon)\rangle + 2\varepsilon \leq \|(f^*, \mu)\| + 2\varepsilon\,.$$

The arbitrariness of $\varepsilon > 0$ implies equality and completes the proof. □

Proof of Proposition 3.11 We fix $f^* \in \mathcal{W}^{1,\infty}(U, \mathbb{R}^m)^*$. With the linear subspace $\tilde{X} := \iota\big(\mathcal{W}^{1,\infty}(U, \mathbb{R}^m)\big)$ of $X := X_0 \times \mathcal{L}^\infty(U, \mathbb{R}^{mn})$, we can use the Hahn-Banach theorem to extend $f^* \circ \iota^{-1} \in \tilde{X}^*$ to some $g^* \in X^*$ under preservation of norm.

Then, by Lemma 3.10, there are $f_0^* \in X_0^*$ and $\mu \in \mathrm{ba}\big(U, \mathcal{B}(U), \mathcal{L}^n\big)^{mn}$ such that

$$\langle f^*, \varphi\rangle = \langle f^* \circ \iota^{-1}, \iota(\varphi)\rangle = \langle g^*, \iota(\varphi)\rangle = \langle f_0^*, \iota_0(\varphi)\rangle + \int_U D\varphi \, d\mu$$

for all $\varphi \in \mathcal{W}^{1,\infty}(U, \mathbb{R}^m)$. Moreover

$$\begin{aligned}\|f_0^*\| + \|\mu\| &= \|(f_0^*, \mu)\| = \|g^*\| = \|f^* \circ \iota^{-1}\| \\ &\le \|f^*\| \, \|\iota^{-1}\| = \|\iota^{-1}\| \sup_{\substack{\varphi \in \mathcal{W}^{1,\infty}(U,\mathbb{R}^m) \\ \|\varphi\|_{\mathcal{W}^{1,\infty}} \le 1}} \langle f^*, \varphi\rangle\end{aligned}$$

which verifies the final estimate. □

Proof of Proposition 3.13 Let $\delta > 0$ be fixed. For $\varphi \in \mathcal{W}^{1,\infty}(U, \mathbb{R}^m)$ we have

$$\chi_\delta^\Gamma \varphi \in \mathcal{W}^{1,\infty}(U, \mathbb{R}^m) \quad \text{and} \quad (1 - \chi_\delta^\Gamma)\varphi|_{\Gamma_{\frac{\delta}{2}}} = 0 \, .$$

Since $f^* \in \mathcal{W}^{1,\infty}(U, \mathbb{R}^m)^*$ is a trace on Γ,

$$\langle f^*, \varphi\rangle = \langle f^*, \chi_\delta^\Gamma \varphi\rangle + \langle f^*, (1 - \chi_\delta^\Gamma)\varphi\rangle = \langle f^*, \chi_\delta^\Gamma \varphi\rangle \, . \tag{3.42}$$

With

$$c_\delta := \|\chi_\delta^\Gamma\|_{\mathcal{W}^{1,\infty}(U,\mathbb{R}^m)} = \|\chi_\delta^\Gamma\|_{\mathcal{W}^{1,\infty}(\Gamma_\delta \cap U,\mathbb{R}^m)} \ge 1$$

and with the product rule for $D(\chi_\delta^\Gamma \varphi)$ we get

$$\begin{aligned}\big\|\chi_\delta^\Gamma \varphi\big\|_{\mathcal{W}^{1,\infty}(U,\mathbb{R}^m)} &= \|\chi_\delta^\Gamma \varphi\|_{\mathcal{W}^{1,\infty}(\Gamma_\delta \cap U,\mathbb{R}^m)} \\ &\le \max\big\{\|\chi_\delta^\Gamma \varphi\|_{\mathcal{L}^\infty(\Gamma_\delta \cap U,\mathbb{R}^m)}, \\ &\qquad \|\chi_\delta^\Gamma D\varphi\|_{\mathcal{L}^\infty(\Gamma_\delta \cap U,\mathbb{R}^m)} + \|\varphi D\chi_\delta^\Gamma\|_{\mathcal{L}^\infty(\Gamma_\delta \cap U,\mathbb{R}^m)}\big\} \\ &\le \max\big\{\|\varphi\|_{\mathcal{L}^\infty(\Gamma_\delta \cap U,\mathbb{R}^m)}, \\ &\qquad \|D\varphi\|_{\mathcal{L}^\infty(\Gamma_\delta \cap U,\mathbb{R}^m)} + c_\delta \|\varphi\|_{\mathcal{L}^\infty(\Gamma_\delta \cap U,\mathbb{R}^m)}\big\} \\ &\le 2c_\delta \max\big\{\|\varphi\|_{\mathcal{L}^\infty(\Gamma_\delta \cap U,\mathbb{R}^m)}, \|D\varphi\|_{\mathcal{L}^\infty(\Gamma_\delta \cap U,\mathbb{R}^m)}\big\} \\ &= 2c_\delta \|\varphi\|_{\mathcal{W}^{1,\infty}(\Gamma_\delta \cap U,\mathbb{R}^m)} \le 2c_\delta \|\varphi\|_{\mathcal{W}^{1,\infty}(U,\mathbb{R}^m)}\end{aligned}$$

for all $\varphi \in \mathcal{W}^{1,\infty}(U, \mathbb{R}^m)$.

We now consider the subspace (that might be strict)

$$X_\delta := \big\{\psi \in \mathcal{W}^{1,\infty}(\Gamma_\delta \cap U, \mathbb{R}^m) \;\big|\; \psi = \varphi_{|\Gamma_\delta \cap U} \;\text{ for some }\; \varphi \in \mathcal{W}^{1,\infty}(U, \mathbb{R}^m)\big\}$$

and define a linear functional f_δ^* on X_δ by

$$\langle f_\delta^*, \varphi_{|\Gamma_\delta \cap U}\rangle = \langle f^*, \chi_\delta^\Gamma \varphi\rangle \quad \text{for} \quad \varphi \in \mathcal{W}^{1,\infty}(U, \mathbb{R}^m)$$

(notice that f_δ^* is well-defined this way). Since

$$\left|\langle f_\delta^*, \varphi_{|\Gamma_\delta \cap U}\rangle\right| = \left|\langle f^*, \chi_\delta^\Gamma \varphi\rangle\right| \le 2c_\delta \|f^*\| \|\varphi\|_{\mathcal{W}^{1,\infty}(\Gamma_\delta \cap U, \mathbb{R}^m)},$$

we have that $f_\delta^* \in X_\delta^*$ and $\|f_\delta^*\| \le 2c_\delta \|f^*\|$. By a norm preserving extension with the Hahn-Banach theorem, we can identify the functional f_δ^* with some functional $f_\delta^* \in \mathcal{W}^{1,\infty}(\Gamma_\delta \cap U, \mathbb{R}^m)^*$. Using (3.42) we obtain

$$\langle f^*, \varphi\rangle = \langle f_\delta^*, \varphi_{|\Gamma_\delta \cap U}\rangle \quad \text{for} \quad \varphi \in \mathcal{W}^{1,\infty}(U, \mathbb{R}^m)$$

which verifies the assertion. □

Proof of Theorem 3.14 For $f^* \in \mathcal{W}^{1,\infty}(U, \mathbb{R}^m)^*$ and $\delta > 0$ we fix

$$f_\delta^* \in \mathcal{W}^{1,\infty}(\Gamma_\delta \cap U, \mathbb{R}^m)^*$$

according to Proposition 3.13. Then we apply Proposition 3.11 with $\Gamma_\delta \cap U$ instead of U, with a suitable choice of X_0, ι_0, and with

$$\|(\varphi, \Phi)\|_{X_0 \times \mathcal{L}^\infty} = \max\{\|\varphi\|_{X_0}, \|\Phi\|_\infty\}.$$

Let us first consider the general case (G) with

$$X_0 = \mathcal{L}^\infty(\Gamma_\delta \cap U, \mathbb{R}^m) \quad \text{and} \quad \iota_0(\varphi) = \varphi.$$

Then

$$\iota : \mathcal{W}^{1,\infty}(\Gamma_\delta \cap U, \mathbb{R}^m) \to \mathcal{L}^\infty(\Gamma_\delta \cap U, \mathbb{R}^m) \times \mathcal{L}^\infty(\Gamma_\delta \cap U, \mathbb{R}^{mn})$$

with $\iota(\varphi) = (\varphi, D\varphi)$ is a linear and isometric mapping. Moreover, it is bijective onto $Y := \iota\big(\mathcal{W}^{1,\infty}(\Gamma_\delta \cap U, \mathbb{R}^m)\big)$. Hence there is a continuous inverse ι^{-1} on Y. With

$$X_0^* = \mathrm{ba}\big(\Gamma_\delta \cap U, \mathcal{B}(\Gamma_\delta \cap U), \mathcal{L}^n\big)^m$$

we obtain the existence of

$$\lambda \in \mathrm{ba}\big(\Gamma_\delta \cap U, \mathcal{B}(\Gamma_\delta \cap U), \mathcal{L}^n\big)^m, \ \mu \in \mathrm{ba}\big(\Gamma_\delta \cap U, \mathcal{B}(\Gamma_\delta \cap U), \mathcal{L}^n\big)^{mn} \tag{3.43}$$

such that the representation of $\langle f^*, \varphi\rangle$ as in (3.34) is satisfied. The measures λ and μ can be extended on U by zero to get (3.33) and, clearly,

$$\operatorname{core}\lambda,\ \operatorname{core}\mu \subset \overline{\Gamma_\delta \cap U}\,.$$

Using (3.30), (3.32), the isometry of ι, and that f_δ^* is a norm preserving extension from the subspace X_δ to X_0 (cf. the proof of Proposition 3.13), we finally have

$$\begin{aligned}
\|\lambda\| + \|\mu\| &= |\lambda|(U) + |\mu|(U) \;=\; |\lambda|(\Gamma_\delta \cap U) + |\mu|(\Gamma_\delta \cap U)\\
&= \|(\lambda,\mu)\| = \|f_\delta^* \circ \iota^{-1}\| = \sup_{\substack{\psi\in W^{1,\infty}(\Gamma_\delta\cap U,\mathbb{R}^m)\\ \|\iota(\psi)\|\le 1}} \langle f_\delta^* \circ \iota^{-1}, \iota(\psi)\rangle\\
&= \sup_{\substack{\psi\in W^{1,\infty}(\Gamma_\delta\cap U,\mathbb{R}^m)\\ \|\psi\|\le 1}} \langle f_\delta^*, \psi\rangle = \sup_{\substack{\psi\in X_\delta\\ \|\psi\|\le 1}} \langle f_\delta^*, \psi\rangle\\
&= \sup_{\substack{\varphi\in W^{1,\infty}(U,\mathbb{R}^m)\\ \|\varphi_{|\Gamma_\delta\cap U}\|\le 1}} \langle f_\delta^*, \varphi_{|\Gamma_\delta\cap U}\rangle = \sup_{\substack{\varphi\in W^{1,\infty}(U,\mathbb{R}^m)\\ \|\varphi_{|\Gamma_\delta\cap U}\|\le 1}} \langle f^*, \chi_\delta^\Gamma \varphi\rangle\,.
\end{aligned}$$

This verifies the first assertion for case (G) (cf. also Adams [1, Theorem 3.8] for the duals of Sobolev spaces).

For case (L) we choose $X_0 = \mathcal{L}_\Gamma^\infty(\Gamma_\delta \cap U, \mathbb{R}^m)$ and $\iota_0(\varphi) = \varphi_{\wr\Gamma}$. Then we combine Proposition 3.11 with Lemma 3.9 to get the existence of measures λ, μ as in (3.43) with $\operatorname{core}\lambda \subset \Gamma$, $\operatorname{core}\mu \subset \overline{\Gamma_\delta \cap U}$, and such that the representation of $\langle f^*, \varphi\rangle$ as in the assertion is true. For (3.33) we extend the measures by zero. With $c = \|\iota^{-1}\| \ge 1$ we get similar to the general case that

$$\begin{aligned}
\|\lambda\| + \|\mu\| &= \|(\lambda,\mu)\| = \|f_\delta^* \circ \iota^{-1}\|\\
&\le \|\iota^{-1}\|\,\|f_\delta^*\| = c \sup_{\substack{\psi\in W^{1,\infty}(\Gamma_\delta\cap U,\mathbb{R}^m)\\ \|\psi\|\le 1}} \langle f_\delta^*, \psi\rangle\\
&= c \sup_{\substack{\varphi\in W^{1,\infty}(U,\mathbb{R}^m)\\ \|\varphi_{|\Gamma_\delta\cap U}\|\le 1}} \langle f^*, \chi_\delta^\Gamma \varphi\rangle\,.
\end{aligned}$$

For case (C) we use $X_0 = C(\Gamma, \mathbb{R}^m)$ and $\iota_0(\varphi) = \varphi_{|\Gamma}$ and then we can argue as in case (L). □

Proof of Proposition 3.15 We first assume that f^* is finite, i.e. (3.37) is satisfied. Then there are $\delta_k > 0$ with $\delta_k \to 0$ and

$$\sup_{k\in\mathbb{N}} \sup_{\substack{\varphi\in W^{1,\infty}(U,\mathbb{R}^m)\\ \|\varphi_{|\Gamma_{\delta_k}\cap U}\|\le 1}} \langle f^*, \chi_{\delta_k}^\Gamma \varphi\rangle < \infty\,. \tag{3.44}$$

Theorem 3.14 provides measures

$$\lambda_k \in \mathrm{ba}\big(U, \mathcal{B}(U), \mathcal{L}^n\big)^m\,, \qquad \mu_k \in \mathrm{ba}\big(U, \mathcal{B}(U), \mathcal{L}^n\big)^{mn}$$

related to δ_k such that $\operatorname{core}\lambda_k, \operatorname{core}\mu_k \subset \overline{\Gamma_{\delta_k} \cap U}$ and, for all $\varphi \in \mathcal{W}^{1,\infty}(U, \mathbb{R}^m)$,

$$\langle f^*, \varphi\rangle = \big\langle (\lambda_k, \mu_k), (\varphi, D_\varphi)\big\rangle = \int_{\Gamma_{\delta_k}\cap U} \varphi\, d\lambda_k + \int_{\Gamma_{\delta_k}\cap U} D\varphi\, d\mu_k\,.$$

By (3.35) for case (G), where c is independent of δ, and by (3.44) there is some $\tilde{c} > 0$ with

$$\|(\lambda_k, \mu_k)\| = \|\lambda_k\| + \|\mu_k\| \le \tilde{c} \quad \text{for all } k\,.$$

Therefore $\{(\lambda_k, \mu_k)\}$ is a bounded sequence in $\big(\mathcal{L}^\infty(U, \mathbb{R}^m) \times \mathcal{L}^\infty(U, \mathbb{R}^{mn})\big)^*$ and, by the Alaoglu theorem, there is a weak* cluster point (λ, μ) with

$$\big\langle (\lambda, \mu), (\varphi, \Phi)\big\rangle = \int_U \varphi\, d\lambda + \int_U \Phi\, d\mu$$

for all $(\varphi, \Phi) \in \mathcal{L}^\infty(U, \mathbb{R}^m) \times \mathcal{L}^\infty(U, \mathbb{R}^{mn})$. Hence there is a subnet of $\{(\lambda_k, \mu_k)\}$ converging to (λ, μ). Thus, for any $(\varphi, \Phi) \in \mathcal{L}^\infty(U, \mathbb{R}^m) \times \mathcal{L}^\infty(U, \mathbb{R}^{mn})$ satisfying

$$\varphi_{|\Gamma_\delta} = 0\,, \quad \Phi_{|\Gamma_\delta} = 0 \quad \text{for some} \quad \delta > 0$$

there is a subsequence $\big\{(\lambda_{k'}, \mu_{k'})\big\}$ such that

$$\begin{aligned} \big\langle (\lambda, \mu), (\varphi, \Phi)\big\rangle &= \lim_{k'\to\infty} \big\langle (\lambda_{k'}, \mu_{k'}), (\varphi, \Phi)\big\rangle \\ &= \lim_{k'\to\infty} \int_U \varphi\, d\lambda_{k'} + \int_U \Phi\, d\mu_{k'} = 0\,. \end{aligned}$$

(recall that $\operatorname{core}\lambda_k, \operatorname{core}\mu_k \subset \Gamma_{\frac{\delta}{2}}$ for k large). Consequently $\operatorname{core}\lambda, \operatorname{core}\mu \subset \Gamma$ and, analogously, $\langle f^*, \varphi\rangle = ⨍_\Gamma \varphi\, d\lambda + ⨍_\Gamma D\varphi\, d\mu$.

For the reverse statement we fix λ, μ as in (3.33) with $\operatorname{core}\lambda, \operatorname{core}\mu \subset \Gamma$ such that for all $\varphi \in \mathcal{W}^{1,\infty}(U, \mathbb{R}^m)$

$$\langle f^*, \varphi\rangle = ⨍_\Gamma \varphi\, d\lambda + ⨍_\Gamma D\varphi\, d\mu\,.$$

Obviously

$$\chi_\delta^\Gamma \varphi = \varphi\,, \quad D(\chi_\delta^\Gamma \varphi) = D\varphi \quad \text{on } \Gamma_{\frac{\delta}{2}} \cap U\,.$$

Thus, for any $\delta > 0$ and any $\varphi \in \mathcal{W}^{1,\infty}(U, \mathbb{R}^m)$ with $\|\varphi_{|\Gamma_\delta \cap U}\|_{\mathcal{W}^{1,\infty}} \le 1$ we use that λ, μ have core in Γ to get

$$\begin{aligned} \langle f^*, \chi_\delta^\Gamma \varphi \rangle &= \fint_\Gamma \chi_\delta^\Gamma \varphi \, d\lambda + \fint_\Gamma D(\chi_\delta^\Gamma \varphi) \, d\mu \\ &= \int_{\Gamma_{\delta/2} \cap U} \varphi \, d\lambda + \int_{\Gamma_{\delta/2} \cap U} D\varphi \, d\mu \\ &\le \|\varphi_{|\Gamma_\delta \cap U}\|_\infty \|\lambda\| + \|(D\varphi)_{|\Gamma_\delta \cap U}\|_\infty \|\mu\| \\ &\le \|\lambda\| + \|\mu\| . \end{aligned}$$

Since the right hand side does not depend on δ, we obtain (3.37).

If we have case (C) for $\delta > 0$, then all $\varphi \in \mathcal{W}^{1,\infty}(U, \mathbb{R}^m)$ can be considered as continuous up to Γ and λ can be replaced by a Radon measure σ with the stated properties according to Proposition 2.13. □

Proof of Proposition 3.16 For (1) we fix $\delta > 0$. First we assume that $\mu = 0$ in (3.34). Then

$$|\langle f^*, \varphi \rangle| = \left| \int_{\Gamma_\delta \cap U} \varphi \, d\lambda \right| \le \|\varphi_{|\Gamma_\delta \cap U}\|_\infty \|\lambda\|$$

for all $\varphi \in \mathcal{W}^{1,\infty}(U, \mathbb{R}^m)$, which verifies the statement. For the other direction we assume (3.38). With f_δ^* from Proposition 3.13 we have

$$|\langle f^*, \varphi \rangle| = \left| \langle f_\delta^*, \varphi_{|\Gamma_\delta \cap U} \rangle \right| \le \tilde{c} \, \|\varphi_{|\Gamma_\delta \cap U}\|_\infty$$

for all $\varphi \in \mathcal{W}^{1,\infty}(U, \mathbb{R}^m)$ and some constant $\tilde{c} > 0$. Hence f_δ^* can be extended to some $g_\delta^* \in \mathcal{L}^\infty(\Gamma_\delta \cap U, \mathbb{R}^m)^*$ by the Hahn-Banach theorem. Consequently there is some measure $\lambda \in \mathrm{ba}\big(U, \mathcal{B}(U), \mathcal{L}^n\big)^m$ with $\|\lambda\| \le \tilde{c}$, $\operatorname{core} \lambda \in \overline{\Gamma_\delta \cap U}$ such that

$$\langle f^*, \varphi \rangle = \int_{\Gamma_\delta \cap U} \varphi \, d\lambda \quad \text{for all} \quad \varphi \in \mathcal{W}^{1,\infty}(U, \mathbb{R}^m) ,$$

which gives the opposite statement. The remaining assertion follows directly from Theorem 3.14.

For (2) we first assume that there is a measure λ with $\operatorname{core} \lambda \subset \Gamma$ and

$$\langle f^*, \varphi \rangle = \fint_\Gamma \varphi \, d\lambda \quad \text{for all} \quad \varphi \in \mathcal{W}^{1,\infty}(U, \mathbb{R}^m) .$$

Then, using (3.32), we have for any $\delta > 0$ and all φ that

$$\langle f^*, \varphi \rangle = \langle f^*, \chi_\delta^\Gamma \varphi \rangle \le \|\varphi_{|\Gamma_\delta \cap U}\|_\infty \|\lambda\| .$$

This readily implies (3.39). For the reverse statement we choose $\delta_k \downarrow 0$ such that the liminf in (3.39) is realized. By the first assertion there are λ_k with $\operatorname{core}\lambda_k \subset \overline{\Gamma_{\delta_k} \cap U}$ and

$$\langle f^*, \varphi \rangle = \int_{\Gamma_{\delta_k} \cap U} \varphi \, d\lambda_k \quad \text{for all} \quad \varphi \in \mathcal{W}^{1,\infty}(U, \mathbb{R}^m)\,.$$

Using the assumption we get for some $\tilde{c} > 0$

$$\sup_{\substack{\varphi \in \mathcal{W}^{1,\infty}(U,\mathbb{R}^m) \\ \|\varphi_{|\Gamma_{\delta_k} \cap U}\|_{\mathcal{L}^\infty} \le 1}} \langle f^*, \varphi \rangle < \tilde{c}$$

for all k. Hence, by the first assertion, we obtain $\|\lambda_k\| \le \tilde{c}$ for all k. Now we can argue as in the proof of Proposition 3.15 to get a weak* cluster point λ of $\{\lambda_k\}$ with $\operatorname{core}\lambda \subset \Gamma$ and

$$\langle f^*, \varphi \rangle = \fint_{\Gamma} \varphi \, d\lambda \quad \text{for all} \quad \varphi \in \mathcal{W}^{1,\infty}(U, \mathbb{R}^m)\,.$$

Notice that we cannot just apply Propositions 3.15 and assertion (1) simultaneously, since λ in (1) might differ from that in the previous proposition due to non-uniqueness.

For case (C) we argue as in the proof of Proposition 3.15. □

Proof of Proposition 3.17 For (1) we assume that there is a component V of $\Gamma_\delta \cap U$ and some $\delta' > 0$ such that $\Gamma_{\delta'} \cap V = \emptyset$. We consider $\varphi_0, \varphi_1 \in \mathcal{W}^{1,\infty}(\Gamma_\delta \cap U)$ with

$$\varphi_0 = 0 \ \text{ on } \ \Gamma_\delta \cap U\,, \qquad \varphi_1 = \begin{cases} 1 \text{ on } V\,, \\ 0 \text{ otherwise}\,. \end{cases}$$

Obviously $\varphi_0 \ne \varphi_1$. But, for cases (L) and (C) with δ, we have

$$\iota_0(\varphi_0) = \iota_0(\varphi_1) = 0 \quad \text{and} \quad D\varphi_0 = D\varphi_1 = 0\,.$$

Hence $\iota(\varphi_0) = \iota(\varphi_1)$. Therefore ι is not injective. Thus both (L) and (C) are not met, which verifies the assertion.

For (2) we notice that $\varphi \in \mathcal{W}^{1,\infty}(\Gamma_\delta \cap U, \mathbb{R}^m)$ is locally Lipschitz continuous with Lipschitz constant $\|D\varphi\|_\infty$. For $\varepsilon > 0$ there are $x \in \Gamma_\delta \cap U$ and $\delta' \in (0, \delta)$ such that

$$\|\varphi\|_{\mathcal{L}^\infty(\Gamma_\delta \cap U)} - \varepsilon \le |\varphi(x)|\,, \quad \|\varphi\|_{\mathcal{L}^\infty(\Gamma_{\delta'} \cap U)} \le \|\varphi\|_\Gamma + \varepsilon\,.$$

Now we find $y \in \Gamma_{\delta'} \cap U$ that can be connected with x within $\Gamma_\delta \cap U$ by a curve of length less than ℓ. Hence, by local Lipschitz continuity,

$$|\varphi(x)| \le |\varphi(y)| + \ell \|D\varphi\|_\infty .$$

Consequently,

$$\|\varphi\|_{\mathcal{L}^\infty(\Gamma_\delta \cap U)} \le \|\varphi\|_\Gamma + \ell\|D\varphi\|_\infty + 2\varepsilon .$$

Since $\varepsilon > 0$ is arbitrary and $a \cdot b \le |a|_1 |b|_\infty$ for $a, b \in \mathbb{R}^2$, we get with $a = (1, l)$

$$\|\varphi\|_\infty \le \|\varphi\|_\Gamma + \ell\|D\varphi\|_\infty \le (1+\ell) \max\left\{\|\varphi\|_\Gamma, \|D\varphi\|_\infty\right\} .$$

Using that $\iota(\varphi) = (\varphi_{\lfloor\Gamma}, D\varphi)$ for case (L) and that the right hand side is larger than $\|D\varphi\|_\infty$, we obtain

$$\|\varphi\|_{\mathcal{W}^{1,\infty}} = \max\left\{\|\varphi\|_\infty, \|D\varphi\|_\infty\right\} \le (1+\ell)\, \|\iota(\varphi)\| .$$

Therefore ι is injective and ι^{-1} is continuous on its image with $\|\iota^{-1}\| \le 1+\ell$. Observing (3.36) we get the assertion (cf. also Remark 3.12).

For (3) we argue basically as for (2). However we use $\|\varphi\|_{C(\Gamma)}$ instead of $\|\varphi\|_\Gamma$, $\iota(\varphi) = (\varphi_{|\Gamma}, D\varphi)$, and we choose $\delta' > 0$ by continuity of φ such that

$$\|\varphi\|_{C(\Gamma \cup (\Gamma_{\delta'} \cap U))} \le \|\varphi\|_{C(\Gamma)} + \varepsilon .$$

Then we can proceed as above. □

Proof of Corollary 3.18 For (1) we fix $\delta > 0$, $x \in (\partial U)_\delta \cap U$, and $\delta' > 0$. Then there is $x' \in \partial U$ with

$$|x - x'| = \operatorname{dist}_{\partial U} x < \delta .$$

Clearly, the open segment (x', x) belongs to $(\partial U)_\delta \cap U$ and there is $y \in (x', x)$ with $|y - x'| < \min\{\delta, \delta'\}$. Then the closed segment $[y, x]$ connects x, y inside $(\partial U)_\delta \cap U$ and has length less than δ. Hence $(\partial U)_\delta \cap U$ is bounded path connected with ∂U and maximal length $\ell = \delta$. Thus we have case (L) for δ by Proposition 3.17 (2) and $c \le 1 + \delta$ in (3.35).

For (2) we argue as in (1) and use that any $\varphi \in \mathcal{W}^{1,\infty}\big((\partial U)_\delta \cap U, \mathbb{R}^m\big)$ can be extended continuously up to ∂U.

For (3) we fix $\delta > 0$ and observe that all $\varphi \in \mathcal{W}^{1,\infty}(\Gamma_\delta \cap U, \mathbb{R}^m)$ are continuous on $\Gamma_\delta \subset U$. For any $x \in \Gamma_\delta$ and any $\delta' > 0$ we clearly find $y \in \Gamma_{\delta'}$ such that the line segment $[y, x]$ has length less than δ and belongs to Γ_δ. Hence $\Gamma_\delta \cap U$ is bounded path connected with Γ and we have case (C) for δ. For the estimate $c \le 1 + \delta$ we can argue as in the proof of assertion (1). □

Chapter 4
Divergence Theorems

We derive general divergence theorems for vector fields in $\mathcal{DM}^1(U)$ by representing corresponding traces with the results of the previous section. As long as nothing else is mentioned, the cases (G), (L) and (C) are taken for $\Gamma = \partial\Omega$. Sometimes we use functions $\varphi \in \mathcal{W}^{1,\infty}(U, \mathbb{R}^m)$ in integrals on the boundary $\partial\Omega$. Here either $\partial\Omega \subset U$ or φ is continuously extendable on $\partial\Omega$ and we identify it with that extension.

4.1 Divergence Measure Fields

For $F \in \mathcal{DM}^1(U)$ and $\Omega \in \mathcal{B}(U)$ we have that TF given by

$$\langle TF, \varphi \rangle := \operatorname{div}(\varphi F)(\Omega) = \int_\Omega \varphi \, d \operatorname{div} F + \int_\Omega F D\varphi \, d\mathcal{L}^n$$

is a trace on $\partial\Omega$ over $\mathcal{W}^{1,\infty}(U)$ according to Theorem 3.3. Then, with Theorem 3.14, we obtain a general Gauss-Green formula for $\mathcal{DM}^1$-vector fields. Notice that we have to choose $m = 1$ for the particular cases (G), (L) and (C) in this section (recall the related definitions of X_0 after Proposition 3.13).

Theorem 4.1 *Let $U \subset \mathbb{R}^n$ be open and bounded, let $\Omega \in \mathcal{B}(U)$, let $\delta > 0$, and assume that $F \in \mathcal{DM}^1(U)$. Then there exist measures*

$$\lambda_F \in \operatorname{ba}\big(U, \mathcal{B}(U), \mathcal{L}^n\big) \quad \textit{and} \quad \mu_F \in \operatorname{ba}\big(U, \mathcal{B}(U), \mathcal{L}^n\big)^n$$

with $\operatorname{core} \lambda_F, \ \operatorname{core} \mu_F \subset \overline{(\partial\Omega)_\delta \cap U}$ *such that*

$$\langle TF, \varphi \rangle = \operatorname{div}(\varphi F)(\Omega) = \int_{(\partial\Omega)_\delta \cap U} \varphi \, d\lambda_F + \int_{(\partial\Omega)_\delta \cap U} D\varphi \, d\mu_F \tag{4.1}$$

F. Schuricht, M. Schönherr, *A Theory of Traces and the Divergence Theorem*,
Lecture Notes in Mathematics 2372, https://doi.org/10.1007/978-3-031-86664-7_4

for all $\varphi \in \mathcal{W}^{1,\infty}(U)$ *with* $T : \mathcal{DM}^1(U) \to \mathcal{W}^{1,\infty}(U)^*$ *from Theorem 3.3. In the particular cases with* $\Gamma = \partial\Omega$ *we have in addition that:*

(L): λ_F, μ_F *can be chosen such that* $\operatorname{core}\lambda_F \subset \partial\Omega$ *and, thus,* (4.1) *becomes*

$$\operatorname{div}(\varphi F)(\Omega) = \fint_{\partial\Omega} \varphi \, d\lambda_F + \int_{(\partial\Omega)_\delta \cap U} D\varphi \, d\mu_F \,.$$

(C): λ_F, μ_F *can be chosen such that* λ_F *corresponds to a Radon measure* σ_F *with* $\operatorname{supp}\sigma_F \subset \partial\Omega$ *and, thus,*

$$\operatorname{div}(\varphi F)(\Omega) = \int_{\partial\Omega} \varphi \, d\sigma_F + \int_{(\partial\Omega)_\delta \cap U} D\varphi \, d\mu_F$$

(where we always identify φ *with its continuous extension on* $\partial\Omega$*).*

We call (λ_F, μ_F), that represents an element of $\mathcal{W}^{1,\infty}(U)^*$, *normal trace* of F near $\partial\Omega$. In contrast to usual Gauss-Green formulas, (4.1) contains a second boundary term depending on $D\varphi$ and both boundary terms depend on a whole neighborhood of the boundary. It turns out that both extensions cannot be omitted in general. Example 4.9 below shows the necessity of the additional boundary term and from Example 4.10 we see that the dependence on δ is needed.

Proof According to Theorem 3.3 we have that TF with

$$\langle TF, \varphi \rangle = \operatorname{div}(\varphi F)(\Omega)$$

is a trace on $\partial\Omega$ over $\mathcal{W}^{1,\infty}(U)$. Then, for each $\delta > 0$, there are measures λ_F and μ_F as in Theorem 3.14 with $\Gamma = \partial\Omega$. □

For any $\Omega \subset \mathbb{R}^n$ we define the (outward) *unit normal field* ν^Ω of Ω to be the gradient of the signed distance function

$$\nu^\Omega := D\big(\operatorname{dist}_\Omega - \operatorname{dist}_{\Omega^c}\big) \quad \mathcal{L}^n\text{-a.e. on } \mathbb{R}^n \,. \tag{4.2}$$

We avoid confusion with our previous notation ν^Ω by saying that ν^Ω is the usual (measure theoretic) unit normal if we take it on $\partial\Omega$ or $\partial_*\Omega$ and otherwise it is the field given above. Notice that

$$|\nu^\Omega| = 1 \quad \mathcal{L}^n\text{-a.e. on } \operatorname{int}\Omega \cup \operatorname{ext}\Omega \quad \text{and} \tag{4.3}$$

$$\nu^\Omega = 0 \quad \mathcal{L}^n\text{-a.e. on } \partial\Omega$$

(use that $\operatorname{dist}_{\partial\Omega}$ is Lipschitz continuous with Lipschitz constant 1, that it is differentiable $\mathcal{L}^n$-a.e. on $\mathbb{R}^n$ with $|D\operatorname{dist}_{\partial\Omega}(x)| = 1$ at points of differentiability outside $\partial\Omega$, since obviously the directional derivative

$$D\,\mathrm{dist}_{\partial\Omega}\left(x;\tfrac{y-x}{|y-x|}\right) = -1$$

if y is a projection of x onto $\partial\Omega$, and recall [22, pp. 235, 130]; cf. also [11, p. 114]). The coarea formula implies that $\mathcal{H}^{n-1}(\partial\Omega_\delta) < \infty$ for $\mathcal{L}^1$-a.e. δ. Thus, by $\partial_*\Omega_\delta \subset \partial\Omega_\delta$ (cf. [32, p. 50]), such sets have finite perimeter and ν^Ω from (4.2) agrees with the measure theoretic outward unit normal $\mathcal{H}^{n-1}$-a.e. on $\partial\Omega_\delta$ (cf. [11, p. 115] for the last statement). These properties certainly justify to speak about a normal field of Ω.

For $\Omega \subset U$ open or closed we easily get some characterization of the trace TF by several limits using the normal field ν^Ω.

Proposition 4.2 *Let $U \subset \mathbb{R}^n$ be open and bounded and assume that $F \in \mathcal{DM}^1(U)$. If $\Omega \subset U$ is open then*

$$\begin{aligned}
\mathrm{div}(\varphi F)(\Omega) &= \lim_{\delta\downarrow 0}\frac{1}{\delta}\int_{(\partial\Omega)_\delta\cap\Omega}\varphi F\cdot\nu^\Omega\,d\mathcal{L}^n\\
&= \lim_{\delta\downarrow 0}\frac{1}{\delta}\int_0^\delta\int_{\partial\Omega_{-\tau}}\varphi F\cdot\nu^\Omega\,d\mathcal{H}^{n-1}\,d\tau\\
&= \operatorname*{ess\,lim}_{\delta\downarrow 0}\int_{\partial^*\Omega_{-\delta}}\varphi F\cdot\nu^\Omega\,d\mathcal{H}^{n-1}
\end{aligned}$$

and if $\Omega \subset U$ is closed then

$$\begin{aligned}
\mathrm{div}(\varphi F)(\Omega) &= \lim_{\delta\downarrow 0}\frac{1}{\delta}\int_{(\partial\Omega)_\delta\cap\Omega^c}\varphi F\cdot\nu^\Omega\,d\mathcal{L}^n\\
&= \lim_{\delta\downarrow 0}\frac{1}{\delta}\int_0^\delta\int_{\partial\Omega_{\tau}}\varphi F\cdot\nu^\Omega\,d\mathcal{H}^{n-1}\,d\tau\\
&= \operatorname*{ess\,lim}_{\delta\downarrow 0}\int_{\partial^*\Omega_{\delta}}\varphi F\cdot\nu^\Omega\,d\mathcal{H}^{n-1}
\end{aligned}$$

for all $\varphi \in \mathcal{W}^{1,\infty}(U)$ where ess lim *denotes the limit up to an $\mathcal{L}^1$-negligible set.*

The results are a simple evaluation of $\mathrm{div}(\varphi_\delta F)(\Omega)$ for suitable φ_δ. The first two equations for closed Ω can be found in Schuricht [38, p. 534] and [39, p. 189] (cf. also Šilhavý [43, p. 449] for a more general version). The corresponding equations for open Ω can be shown exactly the same way. The third equations can be found in Chen-Comi-Torres [11, pp. 117–123]. Here the approximating sets $\Omega_{-\delta}$ or Ω_δ can be replaced by approximating sets from inside or outside with smooth boundary by using a standard smoothing argument (cf. [31, pp. 129, 150]). The examples below show that the limits on the right hand side need not to be related to a measure on $\partial\Omega$ in general. For Ω where $\mathcal{H}^{n-1}(\partial\Omega_\delta)$ is uniformly bounded near $\partial\Omega$ and for suitable F, Proposition 4.30 below provides a "more classical" version without limit on the right hand side and with some density measure near $\partial\Omega$ of the type given in Proposition 3.1. Let us provide a short proof for the convenience of the reader.

Proof Fix $\varphi \in \mathcal{W}^{1,\infty}(U)$. First let Ω be open and consider $\chi_\delta \in \mathcal{W}^{1,\infty}(U)$ with

$$\chi_\delta := \chi_{\Omega^c} + \tfrac{1}{\delta}\chi_{\Omega\setminus\Omega_{-\delta}} \operatorname{dist}_{\Omega_{-\delta}} \quad \text{for} \quad \delta > 0\,.$$

Obviously $\chi_\delta = 1$ on Ω^c, $\chi_\delta = 0$ on $\Omega_{-\delta}$, $\|1-\chi_\delta\|_{\partial\Omega} = 0$, and $\mathcal{L}^n(\Omega \setminus \operatorname{int}\Omega) = 0$. Then, by Corollary 3.5, by $D\chi_\delta = \frac{1}{\delta}\nu^\Omega$ $\mathcal{L}^n$-a.e. on $\operatorname{supp} D\chi_\delta$, and by dominated convergence,

$$\begin{aligned}
\operatorname{div}(\varphi F)(\Omega) &= \operatorname{div}(\chi_\delta \varphi F)(\Omega)\\
&= \int_\Omega \chi_\delta \varphi \, d \operatorname{div} F + \int_\Omega \chi_\delta F \cdot D\varphi \, d\mathcal{L}^n + \int_\Omega \varphi F \cdot D\chi_\delta \, d\mathcal{L}^n\\
&= \lim_{\delta\downarrow 0} \frac{1}{\delta} \int_{(\partial\Omega_\delta)\cap\Omega} \varphi F \cdot \nu^\Omega \, d\mathcal{L}^n\,.
\end{aligned}$$

The second equation follows from the coarea formula. For the third equation we first observe that $\Omega_{-\delta} \subset \operatorname{int}_*(\Omega_{-\delta}) \subset \overline{\Omega_{-\delta}}$ for $\delta > 0$. Thus the sets $\operatorname{int}_*(\Omega_{-\delta})$ are increasing as $\delta \downarrow 0$ and $\bigcup_{\delta>0} \operatorname{int}_*(\Omega_{-\delta}) = \Omega$. Since $\operatorname{div}(\varphi F)$ is a Radon measure on U,

$$\operatorname{div}(\varphi F)(\Omega) = \lim_{\delta\downarrow 0} \operatorname{div}(\varphi F)\big(\operatorname{int}_*(\Omega_{-\delta})\big)\,.$$

Moreover we have that

$$\operatorname{div}(\varphi F)\big(\operatorname{int}_*(\Omega_{-\delta})\big) = \int_{\partial^*\Omega_{-\delta}} \varphi F \cdot \nu^\Omega \, d\mathcal{H}^{n-1} \quad \text{for } \mathcal{L}^1\text{-a.e. } \delta > 0$$

with the measure theoretic normal ν^Ω on $\partial^*\Omega_{-\delta}$ (cf. [20, p. 212], [38, p. 534]). But, for $\mathcal{L}^1$-a.e. $\delta > 0$, we can replace it with ν^Ω from (4.2) by the arguments following (4.3). This readily gives the third equation.

If Ω is closed we have $\Omega \Subset U$ and it is sufficient to show the assertion for φ_c having compact support in U. Since Ω^c is open and since

$$0 = \operatorname{div}(\varphi_c F)(U) = \operatorname{div}(\varphi_c F)(\Omega) + \operatorname{div}(\varphi_c F)(\Omega^c)$$

by Corollary 3.5, we can apply the first assertion to Ω^c to get the results for Ω. □

Let us discuss some situation where the right hand side in Proposition 4.2 can be represented by a Radon measure supported on $\partial\Omega$ and let us give some relation to continuum mechanics. For that we first recall a result from Schuricht [38, p. 537].

Proposition 4.3 *Let $U \subset \mathbb{R}^n$ be open and bounded and identify $F \in \mathcal{DM}^1(U)$ with a fixed representative. Then there is some $h \in \mathcal{L}^1_{\mathrm{loc}}(U)$ such that for any $\Omega \Subset U$ with finite perimeter and $\int_{\partial_*\Omega} h\, d\mathcal{H}^{n-1} < \infty$ one has $\chi_\Omega F \in \mathcal{DM}^1(U)$. Moreover there is some $g_\Omega \in \mathcal{L}^\infty(U, |\mathrm{div}\, F|)$ with values in $[0,1]$ such that, for any $B \in \mathcal{B}(U)$,*

$$\begin{aligned} \mathrm{div}(\varphi\chi_\Omega F)(B) = \int_B g_\Omega \varphi\, d\,\mathrm{div}\, F + \int_B g_\Omega F \cdot D\varphi\, d\mathcal{L}^n \\ - \int_{\partial_*\Omega\cap B} \varphi F \cdot \nu^\Omega\, d\mathcal{H}^{n-1} \end{aligned} \tag{4.4}$$

for all $\varphi \in \mathcal{W}^{1,\infty}(U)$ and where $g_\Omega(x) = \mathrm{dens}^U_x(\Omega)$ whenever $\mathrm{dens}^U_x(\Omega)$ exists (cf. (2.13)).

Remark 4.4

(1) Notice that there is a sequence $\varepsilon_k \downarrow 0$ with $(F * \eta_{\varepsilon_k})(x) \to F(x)$ if $h(x) < \infty$ by the construction of h. For F and Ω as in Proposition 4.3, Theorem 3.3 gives

$$\mathrm{div}(\varphi\chi_\Omega F)(U) = 0 \quad \text{for all} \quad \varphi \in \mathcal{W}^{1,\infty}(U),$$

since we can change φ outside Ω to have compact support in U. Now we apply (4.4) to $B = U$ and use the disjoint decomposition

$$U = \mathrm{int}_*\Omega \cup \partial_*\Omega \cup (\mathrm{ext}_*\Omega \cap U).$$

Since $g_\Omega = 1$ on $\mathrm{int}_*\Omega$ and $g_\Omega = 0$ on $\mathrm{ext}_*\Omega$ and by $\mathcal{L}^n(\partial_*\Omega) = 0$, we get

$$\int_{\mathrm{int}_*\Omega} \varphi\, d\,\mathrm{div}\, F + \int_{\mathrm{int}_*\Omega} F\cdot D\varphi\, d\mathcal{L}^n = \int_{\partial_*\Omega} \varphi F\cdot\nu^\Omega\, d\mathcal{H}^{n-1} - \int_{\partial_*\Omega} g_\Omega\varphi\, d\,\mathrm{div}\, F\,.$$

For $\Omega = \mathrm{int}_*\Omega$ we thus have that $\mathrm{div}(\varphi F)(\Omega)$ is related to a Radon measure supported on $\partial_*\Omega$. Notice that the result covers cases where $\mathrm{div}\, F$ does not vanish on $\partial\Omega$. If $F \in \mathcal{DM}^\infty(U)$ this is true with $g_\Omega = \frac{1}{2}$, since $\mathrm{div}\, F <<^w \mathcal{H}^{n-1}$ and $g_\Omega = \frac{1}{2}$ for $\mathcal{H}^{n-1}$-a.e. point on $\partial_*\Omega$, and the related Radon measure has an $\mathcal{H}^{n-1}$-integrable density on $\partial_*\Omega$ (cf. [2, p. 14]).

(2) The Gauss-Green formula plays also an important role for contact interactions in continuum mechanics. Let U be related to a continuous body, let F and $\Omega = \mathrm{int}_*\Omega$ be as in Proposition 4.3, and let $B \subset (\mathrm{int}_*\Omega)^c$ be a Borel set. Then, using $g_\Omega = 0$ on $\mathrm{ext}_*\Omega$, we directly get

$$\mathrm{div}(\chi_\Omega F)(B) = \int_{\partial_*\Omega\cap B} g_\Omega\, d\,\mathrm{div}\, F - \int_{\partial_*\Omega\cap B} F\cdot\nu^\Omega\, d\mathcal{H}^{n-1}.$$

This gives the action exerted from the subbody related to Ω to the subbody related to B. Analogously as above, this covers cases where $\mathrm{div}\, F$ does not vanish on $\partial\Omega$ and we can specialize it for essentially bounded F (cf. Schuricht [38, p. 536], Chen-Torres-Ziemer [10, pp. 291, 298], Chen-Comi-Torres [11, p. 157]).

Let us now characterize the special cases where $\mu_F = 0$ is possible and where the measures λ_F, μ_F can be chosen independent of δ in (4.1).

Proposition 4.5 *Let $U \subset \mathbb{R}^n$ be open and bounded, let $\Omega \in \mathcal{B}(U)$, and assume that $F \in \mathcal{DM}^1(U)$.*

(1) *In Theorem* 4.1 *we can choose (λ_F, μ_F) with* $\operatorname{core}\lambda_F$, $\operatorname{core}\mu_F \subset \partial\Omega$, *i.e. λ_F, μ_F are independent of δ, if and only if*

$$\liminf_{\delta\downarrow 0} \sup_{\substack{\varphi\in\mathcal{W}^{1,\infty}(U)\\ \|\varphi_{|(\partial\Omega)_\delta\cap U}\|_{\mathcal{W}^{1,\infty}}\le 1}} \operatorname{div}(\chi_\delta^{\partial\Omega}\varphi F)(\Omega) < \infty \tag{4.5}$$

with $\chi_\delta^{\partial\Omega}$ as in (3.31). *In this case* (4.1) *becomes*

$$\operatorname{div}(\varphi F)(\Omega) = \fint_{\partial\Omega} \varphi\, d\lambda_F + \fint_{\partial\Omega} D\varphi\, d\mu_F\,. \tag{4.6}$$

(2) *In Theorem* 4.1 *we can choose (λ_F, μ_F) for $\delta > 0$ such that $\mu_F = 0$ and* $\operatorname{core}\lambda_F \subset \overline{(\partial\Omega)_\delta \cap U}$ *if and only if*

$$\sup_{\substack{\varphi\in\mathcal{W}^{1,\infty}(U)\\ \|\varphi_{|(\partial\Omega)_\delta\cap U}\|_{\mathcal{L}^\infty}\le 1}} \operatorname{div}(\varphi F)(\Omega) < \infty\,. \tag{4.7}$$

(3) *In Theorem* 4.1 *we can take (λ_F, μ_F) with $\mu_F = 0$ and* $\operatorname{core}\lambda_F \subset \partial\Omega$ *if and only if*

$$\liminf_{\delta\downarrow 0} \sup_{\substack{\varphi\in\mathcal{W}^{1,\infty}(U)\\ \|\varphi_{|(\partial\Omega)_\delta\cap U}\|_{\mathcal{L}^\infty}\le 1}} \operatorname{div}(\varphi F)(\Omega) < \infty\,. \tag{4.8}$$

Notice that (4.5) just means that the trace functional $\varphi \to \operatorname{div}(\varphi F)(\Omega)$ is finite.

Proof Let TF be as in Theorem 4.1. Then (1) is a direct consequence of Proposition 3.15 and (2), (3) follow directly from Proposition 3.16. □

In addition, we provide some equivalent conditions for (4.5).

Lemma 4.6 *Let $U \subset \mathbb{R}^n$ be open and bounded, let $\Omega \in \mathcal{B}(U)$, and assume that $F \in \mathcal{DM}^1(U)$. Then* (4.5) *is equivalent to each of the following two conditions:*

(1)

$$\liminf_{\delta\downarrow 0} \sup_{\substack{\varphi\in\mathcal{W}^{1,\infty}(U)\\ \|\varphi_{|(\partial\Omega)_\delta\cap U}\|_{\mathcal{W}^{1,\infty}}\le 1}} \int_{(\partial\Omega)_\delta\cap\Omega} \varphi F D\chi_\delta^{\partial\Omega}\, d\mathcal{L}^n < \infty\,, \tag{4.9}$$

(2)

$$\liminf_{\delta\downarrow 0} \sup_{\substack{\varphi\in\mathcal{W}^{1,\infty}(U)\\ \|\varphi_{|(\partial\Omega)_\delta\cap U}\|_{\mathcal{W}^{1,\infty}}\le 1}} \frac{1}{\delta}\int_{((\partial\Omega)_\delta\setminus(\partial\Omega)_{\frac{\delta}{2}})\cap\Omega} \varphi F D\mathrm{dist}_{\partial\Omega}\, d\mathcal{L}^n < \infty\,.$$

Moreover, $\operatorname{int}\Omega=\emptyset$ *implies* (4.5).

Proof By (3.21) and (3.22) we have for $\varphi\in\mathcal{W}^{1,\infty}(U)$ that

$$\operatorname{div}(\chi_\delta^{\partial\Omega}\varphi F) = \chi_\delta^{\partial\Omega}\varphi\operatorname{div}F + \chi_\delta^{\partial\Omega}FD\varphi\mathcal{L}^n + \varphi FD\chi_\delta^{\partial\Omega}\mathcal{L}^n$$

as measures on U. Obviously

$$\left|\chi_\delta^{\partial\Omega}\varphi\operatorname{div}F\right|(\Omega) \quad\text{and}\quad \left|\chi_\delta^{\partial\Omega}FD\varphi\mathcal{L}^n\right|(\Omega)$$

are uniformly bounded for $\delta>0$ and $\|\varphi_{|(\partial\Omega)_\delta\cap U}\|_{\mathcal{W}^{1,\infty}}\le 1$. Since

$$(\varphi FD\chi_\delta^{\partial\Omega}\mathcal{L}^n)(\Omega) = \int_{(\partial\Omega)_\delta\cap\Omega}\varphi FD\chi_\delta^{\partial\Omega}\,d\mathcal{L}^n\,,$$

(4.5) is equivalent to (4.9). For the second condition we use that

$$D\chi_\delta^{\partial\Omega} = -\tfrac{2}{\delta}\chi_{(\partial\Omega)_\delta\setminus(\partial\Omega)_{\delta/2}}D\,\mathrm{dist}_{\partial\Omega}\,.$$

If $\operatorname{int}\Omega=\emptyset$, then $D\chi_\delta^{\partial\Omega}=0$ for all x and this readily implies (4.9). □

Let us also give some sufficient conditions that are useful for applications.

Proposition 4.7 *Let* $U\subset\mathbb{R}^n$ *be an open and bounded set, let* $\Omega\in\mathcal{B}(U)$ *with* $\mathcal{L}^n(\Omega\setminus\operatorname{int}\Omega)=0$, *and assume that* $F\in\mathcal{DM}^1(U)$. *If*

$$\liminf_{\delta\downarrow 0}\int_\Omega \left|FD\chi_\delta^{\partial\Omega}\right| d\mathcal{L}^n < \infty\,, \tag{4.10}$$

then we can choose $\mu_F=0$ *and* λ_F *with* $\operatorname{core}\lambda_F\subset\partial\Omega$ *in Theorem* 4.1. *We have* (4.10) *if* F *is bounded and if there is some* $\tilde\delta>0$ *such that*

$$\sup_{\delta\in(0,\tilde\delta)}\mathcal{H}^{n-1}(\partial\Omega_{-\delta})<\infty\,. \tag{4.11}$$

If Ω has finite perimeter and if there are $c > 0$, $r > 0$ such that

$$\frac{\mathcal{L}^n(B_\delta(x) \cap (\overline{\Omega})^c)}{\mathcal{L}^n(B_\delta(x))} \geq c \quad \textit{for all} \quad x \in \partial\Omega\,,\ \delta \in (0, r)\,, \tag{4.12}$$

then (4.11) *is satisfied.*

Notice that $\partial\Omega_{-\delta}$ has finite perimeter for all $\delta > 0$ (cf. [29, p. 2788]. However, even if Ω has finite perimeter, the perimeters $\mathcal{H}^{n-1}(\partial\Omega_{-\delta})$ might not be bounded uniformly in δ (cf. [29, p. 2781]). The sufficient condition (4.12) is a uniform lower bound for the density of the exterior of Ω at $x \in \partial\Omega$. It in particular excludes points $x \in \partial\Omega$ where this density vanishes (as, e.g., for x at an inward cusp of Ω or for inner boundary points $x \in \operatorname{int}\overline{\Omega}$). But (4.12) is obviously met if Ω has Lipschitz boundary. In this case we can also continuously extend all $\varphi \in \mathcal{W}^{1,\infty}(U)$ up to $\partial\Omega$ such that λ_F can be considered as Radon measure supported on $\partial\Omega$. Let us also refer to [43, p. 449] where it is shown that a condition similar to (4.10) allows the representation of $\operatorname{div}(\varphi F)(\Omega)$ by a σ-measure on $\partial\Omega$ for open and bounded Ω and φ that are Lipschitz continuous on $\mathbb{R}^n$.

Proof By (4.10) there is some $c > 0$ and a sequence $\delta_k \downarrow 0$ such that

$$\int_\Omega \left|FD\chi_{\delta_k}^{\partial\Omega}\right| d\mathcal{L}^n \leq c \quad \text{for all} \quad k\,.$$

For the trace TF from Theorem 4.1 and for $\varphi \in \mathcal{W}^{1,\infty}(U)$ we can use dominated convergence to get

$$\begin{aligned}
\langle TF, \varphi\rangle = \operatorname{div}(\varphi F)(\Omega) &= \lim_{k\to\infty} \operatorname{div}(\chi_{\delta_k}^{\partial\Omega}\varphi F)(\Omega) \\
&= \lim_{k\to\infty} \int_\Omega \chi_{\delta_k}^{\partial\Omega}\varphi \, d\operatorname{div} F + \int_\Omega \chi_{\delta_k}^{\partial\Omega} FD\varphi \, d\mathcal{L}^n + \int_\Omega \varphi F D\chi_{\delta_k}^{\partial\Omega} \, d\mathcal{L}^n \\
&\leq \lim_{k\to\infty} \big(|\operatorname{div} F|(\partial\Omega) + c\big) \, \|\varphi_{|(\partial\Omega)_{\delta_k}\cap U}\|_\infty\,.
\end{aligned}$$

Hence the first statement follows from Proposition 4.5 (3). For bounded F the coarea formula implies

$$\int_\Omega \left|FD\chi_\delta^{\partial\Omega}\right| d\mathcal{L}^n \leq \|F\|_\infty \int_\Omega \left|D\chi_\delta^{\partial\Omega}\right| d\mathcal{L}^n = \tfrac{2}{\delta}\|F\|_\infty \int_{-\delta}^{-\frac{\delta}{2}} \mathcal{H}^{n-1}(\partial\Omega_\tau)\, d\tau$$

which gives the second assertion. For the last statement, (4.12) just means that the open set $(\overline{\Omega})^c$ has (r, c)-uniform lower density on $\partial\Omega$ in the sense of Definition 4 in [29]. Hence, by Theorems 3 and 4 in [29], there are constants $c_1, c_2 > 0$ such that for all $\delta \in (0, r)$ and with Per denoting the perimeter

$$\mathcal{H}^{n-1}(\partial\Omega_\delta) \le c_1 \frac{\mathcal{L}^n(\Omega \setminus \Omega_\delta)}{\delta} \le c_2 \,\mathrm{Per}\left((\overline{\Omega})^c\right). \tag{4.13}$$

Since $\mathrm{Per}\left((\overline{\Omega})^c\right) = \mathrm{Per}(\Omega)$, we readily get (4.11). □

For Lipschitz continuous functions $\varphi \in \mathrm{Lip}(\Gamma)$ with $\Gamma \subset \mathbb{R}^n$ we use the norm

$$\|\varphi\|_{\mathrm{Lip}(\Gamma)} = \|\varphi\|_{C(\Gamma)} + \mathrm{Lip}(\varphi)$$

where $\mathrm{Lip}(\varphi)$ is the (smallest) Lipschitz constant of φ on Γ.

Proposition 4.8 *Let $F \in \mathcal{DM}^1(U)$ for some open bounded $U \subset \mathbb{R}^n$.*

(1) *If $\Omega \in \mathcal{B}(U)$ is such that any $\varphi \in \mathcal{W}^{1,\infty}(U)$ has a continuous extension onto $\overline{\Omega}$, if $\mathcal{L}^n(\Omega \setminus \mathrm{int}\,\Omega) = 0$, and if there are $c > 0$ and $\tilde{\delta} > 0$ such that*

$$\|\varphi_{|\partial\Omega}\|_{\mathrm{Lip}(\partial\Omega)} \le c\|\varphi\|_{\mathcal{W}^{1,\infty}((\partial\Omega)_\delta \cap U)} \quad \text{for all} \quad \varphi \in \mathcal{W}^{1,\infty}(U),\ \delta \in (0, \tilde{\delta}), \tag{4.14}$$

then (4.5) *is satisfied.*

(2) *If $\Omega \subset U$ is open with Lipschitz boundary, then* (4.5) *is satisfied.*

Notice that (4.14) is a condition for Ω that does not depend on F. It somehow says that $\mathcal{W}^{1,\infty}$-functions should be (globally) Lipschitz continuous near the boundary. Since the proof of Proposition 4.8 is quite technical, we postpone it to the end of this section and now provide several examples which illustrate the previous results. First we show by an example of some open $\Omega \subset \mathbb{R}^2$ with Lipschitz boundary and some unbounded F that $\mu_F = 0$ in (4.1) is not possible in general. This way we see that the new term is really needed for a general Gauss-Green formula.

Example 4.9 Let $U = B_2(0) \subset \mathbb{R}^2$, let $\Omega = (0, 1)^2$, and let $F \in \mathcal{L}^1(U, \mathbb{R}^2)$ be given by

$$F(x, y) := \frac{1}{x^2 + y^2}\begin{pmatrix} -y \\ x \end{pmatrix}.$$

(cf. also [43, Example 2.5]). We clearly have $\mathrm{div}\, F = 0$ on $U \setminus \{0\}$ in the classical sense and $F \cdot \nu^{B_\delta(0)} = 0$ on $\partial B_\delta(0)$ for $\delta > 0$. Then, for all $\varphi \in C_c^1(U)$,

$$\begin{aligned} \int_U F D\varphi \, d\mathcal{L}^2 &= \lim_{\delta \downarrow 0} \int_{U \setminus B_\delta(0)} F D\varphi \, d\mathcal{L}^2 \\ &= -\lim_{\delta \downarrow 0} \left(\int_{U \setminus B_\delta(0)} \varphi \,\mathrm{div}\, F \, d\mathcal{L}^2 + \int_{\partial B_\delta(0)} \varphi F \cdot \nu^{B_\delta(0)} \, d\mathcal{H}^1 \right) \\ &= 0\,. \end{aligned}$$

Therefore div F is the zero measure on U and thus, $F \in \mathcal{DM}^1(U)$. For small $\delta > 0$ we have case (C) by Corollary 3.18 (3) and (4.5) is satisfied by Proposition 4.8 (2). Hence, by Theorem 4.1, there are a Radon measure $\sigma_F \in \mathcal{M}(U)$ supported on $\partial\Omega$ and a measure $\mu_F \in \mathrm{ba}(U, \mathcal{B}(U), \mathcal{L}^2)^2$ with core in $\partial\Omega$ such that

$$\operatorname{div}(\varphi F)(\Omega) = \int_{\partial\Omega} \varphi \, d\sigma_F + \int_{\partial\Omega} D\varphi \, d\mu_F \tag{4.15}$$

for all $\varphi \in \mathcal{W}^{1,\infty}(U)$. For

$$\varphi_k := \chi_{(\frac{1}{k},\infty)\times\mathbb{R}} + \chi_{(0,\frac{1}{k})\times\mathbb{R}} \, k \operatorname{dist}_{\{0\}\times\mathbb{R}}$$

we have

$$\varphi_k \in \mathcal{W}^{1,\infty}(U), \quad D\varphi_k = \chi_{(0,\frac{1}{k})\times\mathbb{R}} \, k \begin{pmatrix} 1 \\ 0 \end{pmatrix}, \quad 0 \le \varphi_k \le 1$$

and thus,

$$\begin{aligned}
\big|\operatorname{div}(\varphi_k F)(\Omega)\big| &= \left| \int_\Omega F D\varphi_k \, d\mathcal{L}^2 + \int_\Omega \varphi_k \, d \operatorname{div} F \right| = \left| \int_\Omega F D\varphi_k \, d\mathcal{L}^2 \right| \\
&= \int_{(0,\frac{1}{k})} \int_{(0,1)} \frac{y}{x^2+y^2} \, dy \, dx = \int_{(0,\frac{1}{k})} \left[\frac{1}{2} \ln(x^2+y^2) \right]_{y=0}^{1} dx \\
&= \int_{(0,\frac{1}{k})} \frac{1}{2} \ln\left(\frac{1}{x^2} + 1 \right) dx \\
&\ge \frac{1}{2} \ln(k^2+1) \xrightarrow{k\to\infty} \infty .
\end{aligned}$$

Moreover

$$\left| \int_{\partial\Omega} \varphi_k \, d\sigma_F \right| \le |\sigma_F|(U)$$

for all k. Hence $\mu_F = 0$ is impossible in (4.15).

In Example 4.17 below we consider the same vector field on a slightly modified set Ω. There we construct for (4.15) possible Radon measures σ_F and pure measures μ_F depending on some scalar parameter. This way we provide an uncountable family of possibilities for σ_F and μ_F.

Next we provide an example with a constant vector field F and an open $\Omega \subset \mathbb{R}^2$ having infinite perimeter where (4.5) fails. This means by Proposition 4.5 that the dependence of the measures λ_F, μ_F on $\delta > 0$ cannot be removed.

Example 4.10 Let $U := B_2(0) \subset \mathbb{R}^2$, let

$$\Omega := \bigcup_{k=1}^{\infty} R_k \quad \text{with} \quad R_k := \left(\tfrac{1}{2k+1}, \tfrac{1}{2k}\right) \times (0, 1),$$

and take the constant vector field

$$F \in \mathcal{L}^1(U, \mathbb{R}^2) \quad \text{with} \quad F = \left(\begin{smallmatrix}1\\0\end{smallmatrix}\right).$$

(cf. also Example 3.20). Obviously $\operatorname{div} F = 0$ on U and thus $F \in \mathcal{DM}^1(U)$.

Let us show that (4.9) does not hold. For $\delta > 0$ we choose $k_\delta \in \mathbb{N}$ to be the largest number such that

$$\delta < \tfrac{1}{4}\left(\tfrac{1}{2k} - \tfrac{1}{2k+1}\right) \quad \text{for all} \quad k \le k_\delta + 1 .$$

Then $k_\delta \to \infty$ for $\delta \to 0$. Moreover, for $\delta > 0$ fixed, we set

$$R_k^- := \left(\tfrac{1}{2k+1}, \tfrac{1}{2k+1} + \delta\right) \times (0, 1), \quad R_k^+ := \left(\tfrac{1}{2k} - \delta, \tfrac{1}{2k}\right) \times (0, 1).$$

Then we obviously find some

$$\varphi^\delta \in \mathcal{W}^{1,\infty}(U) \quad \text{with} \quad \|\varphi^\delta_{|(\partial\Omega)_\delta \cap U}\|_{\mathcal{W}^{1,\infty}} \le 1$$

such that

$$\varphi^\delta(x, y) = \begin{cases} \pm\frac{1}{2}\left(\frac{1}{2} - |y - \frac{1}{2}|\right) & \text{on } R_k^\pm \text{ for } k \le k_\delta , \\ 0 & \text{on } R_k \text{ for } k > k_\delta . \end{cases}$$

Notice that

$$\varphi^\delta(x, 0) = \varphi^\delta(x, 1) = 0, \quad \varphi^\delta(x, \tfrac{1}{2}) = \pm\tfrac{1}{4} \quad \text{on } R_k^\pm \text{ for } k \le k_\delta ,$$

$$|D\varphi^\delta(x, y)| = \tfrac{1}{2} \quad \text{on } R_k^\pm \text{ for } k \le k_\delta ,$$

$$F D\chi_\delta^{\partial\Omega} = 0 \quad \text{on} \quad \left((\partial\Omega)_\delta \cap \Omega\right) \setminus \left(R_k^- \cup R_k^+\right).$$

Hence

$$\begin{aligned}\int_{(\partial\Omega)_\delta\cap\Omega} \varphi^\delta F D\chi_\delta^{\partial\Omega}\, d\mathcal{L}^2 &= \sum_{k=1}^{k_\delta} \int_{R_k^-\cup R_k^+} \varphi^\delta F D\chi_\delta^{\partial\Omega}\, d\mathcal{L}^2 \\ &\geq \sum_{k=1}^{k_\delta} 2 \int_\delta^{1-\delta} \tfrac{1}{2}\big(\tfrac{1}{2} - |y-\tfrac{1}{2}|\big)\, dy \\ &= 2k_\delta \int_\delta^{\frac{1}{2}} y\, dy \;=\; k_\delta\big(\tfrac{1}{4}-\delta^2\big) \overset{\delta\to 0}{\longrightarrow} \infty .\end{aligned}$$

But this means that (4.9) is not satisfied.

For $\delta \in \big(0, \frac{1}{2}\big)$ and $\Gamma = \partial\Omega$ we have case (C) by Corollary 3.18 (3) and (4.1) becomes

$$\operatorname{div}(\varphi F)(\Omega) = \int_{\partial\Omega} \varphi\, d\sigma_F + \int_{(\partial\Omega)_\delta\cap U} D\varphi\, d\mu_F$$

(notice that this does not contradict Example 3.20 where U and Γ are different).

Let us provide possible choices of σ_F and μ_F. For $\delta > 0$ we first fix some $m_\delta \in \mathbb{N}$ such that

$$\frac{1}{2m_\delta} - \frac{1}{2m_\delta + 1} < \delta .$$

Then we get for every $\varphi \in \mathcal{W}^{1,\infty}(U)$

$$\begin{aligned}\operatorname{div}(\varphi F)(\Omega) = \sum_{k=1}^{\infty} \operatorname{div}(\varphi F)(R_k) \;&=\; \sum_{k=1}^{\infty} \int_{R_k} \varphi \operatorname{div} F + F D\varphi\, d\mathcal{L}^2 \\ &= \sum_{k=1}^{m_\delta} \int_{\partial R_k} \varphi F \nu^{R_k}\, d\mathcal{H}^1 + \sum_{k=m_\delta+1}^{\infty} \int_{R_k} F D\varphi\, d\mathcal{L}^2\end{aligned}$$

where we have used the classical Gauss-Green formula for $k \leq m_\delta$. Hence we can choose

$$\sigma_F^\delta = \sum_{k=1}^{m_\delta} F\nu^{R_k}\mathcal{H}^1\llcorner\partial R_k\,, \quad \mu_F^\delta = \sum_{k=m_\delta+1}^{\infty} F\mathcal{L}^2\llcorner R_k\,.$$

Notice that $R_k \subset (\partial\Omega)_\delta \cap U$ for $k \geq m_\delta + 1$ and that the measures μ_F^δ are even σ-measures. Let us also mention that we cannot sum over all $k \in \mathbb{N}$ for σ_F, since this would not give a bounded measure. The dependence of the measures on δ comes through m_δ. But, for fixed $\delta > 0$, we also have some freedom to choose $m_\delta \in \mathbb{N}$. Therefore the choice of σ_F and μ_F is not unique even for given δ. Since the measures σ_F^δ, μ_F^δ given above are restricted to Ω, the situation would not change if we take $U = \Omega$ instead of $U = B_2(0)$.

We now give an example where we can choose $\mu_F = 0$ but λ_F cannot be taken as σ-measure.

Example 4.11 We set

$$\Omega_1 := (0, 1) \times (0, 1), \quad \Omega_2 := (1, 2) \times (0, 1)$$

and consider

$$U = \Omega = \Omega_1 \cup \Omega_2$$

with the discontinuous vector field

$$F = \begin{pmatrix} 1 \\ 0 \end{pmatrix} \text{ on } \Omega_1, \quad F = \begin{pmatrix} -1 \\ 0 \end{pmatrix} \text{ on } \Omega_2.$$

We readily verify (4.11) and hence we can choose $\mu_F = 0$ and λ_F with $\operatorname{core}\lambda_F \subset \partial\Omega$ in Theorem 4.1. Since $\varphi \in \mathcal{W}^{1,\infty}(U)$ cannot be extended continuously up to $\partial\Omega$ in general, we do not have case (C) and we cannot expect that λ_F is in fact a Radon measure on $\partial\Omega$. But notice that we have case (L) by Corollary 3.18 (1).

Though the classical Gauss-Green formula is not applicable on Ω, we can use it on Ω_j, since $\varphi_{|\Omega_j}$ is continuously extendable to a Lipschitz function $\overline{\varphi}_j$ on $\overline{\Omega}_j$ for all $\varphi \in \mathcal{W}^{1,\infty}(U)$. Therefore, with $\sigma_F^j = F\nu^{\Omega_j}\mathcal{H}^1 \lfloor \partial\Omega_j$,

$$\operatorname{div}(\varphi F)(\Omega_j) = \int_{\partial\Omega_j} \overline{\varphi}_j F\nu^{\Omega_j}\, d\mathcal{H}^1 = \int_{\partial\Omega_j} \overline{\varphi}_j\, d\sigma_F^j \quad \text{for} \quad \varphi \in \mathcal{W}^{1,\infty}(U).$$

But since φ does not have a continuous extension up to $\partial\Omega$ in general, we cannot just sum up the σ_F^j for Ω. However we can apply Proposition 2.13 (2) to Ω_j and obtain pure measures

$$\lambda_F^j \in \operatorname{ba}(\Omega_j, \mathcal{B}(\Omega_j), \mathcal{L}^2) \quad \text{with} \quad \operatorname{core}\lambda_F^j = \partial\Omega_j$$

such that

$$\operatorname{div}(\varphi F)(\Omega_j) = \fint_{\partial\Omega_j} \varphi_{|\Omega_j}\, d\lambda_F^j \quad \text{for} \quad \varphi \in \mathcal{W}^{1,\infty}(U).$$

We can consider the λ_F^j as measures on Ω by extending them by zero. Then Ω_j is an aura of λ_F^j and we readily obtain for $\lambda_F = \lambda_F^1 + \lambda_F^2$ that

$$\operatorname{div}(\varphi F)(\Omega) = \fint_{\partial\Omega} \varphi\, d\lambda_F \quad \text{for} \quad \varphi \in \mathcal{W}^{1,\infty}(U).$$

We can interpret the situation along the common boundary of the Ω_j so, that one part of λ_F takes care of $\varphi_{|\Omega_1}$ and the other part takes care of $\varphi_{|\Omega_2}$. This nicely shows the relevance of the aura. In the case of a crack along the inner boundary we would be able to describe the situation on each side of the crack separately by choosing suitable functions φ.

Notice that the usual Gauss-Green formula using $\mathrm{int}_*\Omega$ and $\partial^*\Omega$ (cf. [22, p. 209]) is substantially different, since here the interior part of $\partial\Omega$ belongs to $\mathrm{int}_*\Omega$ and φ has to be continuous on $\mathrm{int}_*\Omega$.

In the next example we consider some vector field F where div F has some point concentration on $\partial\Omega$. We demonstrate the difference between an open Ω and its closure and we discuss some lower dimensional Ω.

Example 4.12 We consider the vector field, that is a classical example in the literature (cf. e.g. Chen-Frid [6, p. 403]),

$$F(x) := \frac{x}{2\pi |x|^2} \quad \text{on} \quad U = B_2(0) \subset \mathbb{R}^2$$

and we discuss the Gauss-Green formula for several Ω. Clearly $F \in \mathcal{L}^1(U, \mathbb{R}^2)$ and we easily compute that $\operatorname{div} F = 0$ on $U \setminus \{0\}$ in the classical sense. Moreover, with the usual Gauss-Green formula we get for $\varphi \in C_c^1(U)$

$$\begin{aligned} \int_U F D\varphi \, d\mathcal{L}^2 &= \lim_{\delta \downarrow 0} \int_{U \setminus B_\delta(0)} F D\varphi \, d\mathcal{L}^2 \\ &= -\lim_{\delta \downarrow 0} \Big(\int_{U \setminus B_\delta(0)} \varphi \operatorname{div} F \, d\mathcal{L}^2 + \int_{\partial B_\delta(0)} \varphi F \cdot \nu^{B_\delta(0)} \, d\mathcal{H}^1 \Big) \\ &= -\lim_{\delta \downarrow 0} \frac{1}{2\pi\delta} \int_{\partial B_\delta(0)} \varphi \, d\mathcal{H}^1 \ = -\varphi(0) \, . \end{aligned}$$

Hence, by (3.14), $\operatorname{div} F = \delta_0$ as measure (Dirac measure concentrated at the origin) and $F \in \mathcal{DM}^1(U)$.

First we consider

$$\Omega := B_1(0) \cap C_\alpha \subset \mathbb{R}^2$$

where C_α is an open cone with opening angle $\alpha \in (0, 2\pi)$ and with vertex at the origin. Let us check condition (4.10). There we have to integrate over the support of $\chi_\delta^{\partial\Omega}$. For the integral over the curved part near $\partial B_1(0)$ we use the coarea formula to get a bound for small $\delta > 0$ by

$$\int_{(\partial B_1(0))_\delta \cap B_1(0)} \big| F D\chi_\delta^{\partial\Omega} \big| \, d\mathcal{L}^2 = \frac{2}{2\pi\delta} \int_{1-\delta}^{1-\frac{\delta}{2}} \frac{2\pi r}{r} \, dr = 1 \, .$$

For ∂C_α we assume that it contains the positive x_1-axis. Then, for small $\delta > 0$,

$$\Omega \cap Q_\delta \neq \emptyset \quad \text{for} \quad Q_\delta = \left(\tfrac{\delta}{2}, 1\right) \times \left(\tfrac{\delta}{2}, \delta\right).$$

Using the symmetry, we can estimate

$$\begin{aligned}
\int_\Omega \left|F D\chi_\delta^{\partial\Omega}\right| d\mathcal{L}^2 &\leq 1 + 2\int_{Q_\delta} \frac{(x_1, x_2)}{2\pi(x_1^2 + x_2^2)} \cdot \left(0, \tfrac{2}{\delta}\right) d\mathcal{L}^2 \\
&= 1 + \frac{2}{\delta\pi} \int_{\frac{\delta}{2}}^{\delta} \int_{\frac{\delta}{2}}^{1} \frac{x_2}{x_1^2 + x_2^2}\, dx_1 dx_2 \\
&\leq 1 + \frac{2}{\delta\pi}\frac{\delta}{2} \int_{\frac{\delta}{2}}^{1} \frac{\delta}{x_1^2 + \frac{\delta^2}{4}}\, dx_1 \;=\; 1 + \frac{2}{\pi}\left[\arctan \tfrac{2x_1}{\delta}\right]_{\delta/2}^{1} \\
&= 1 + \tfrac{2}{\pi}\left(\arctan \tfrac{2}{\delta} - \arctan 1\right).
\end{aligned}$$

Hence (4.10) is satisfied. Thus we can choose $\mu_F = 0$ and λ_F with $\operatorname{core} \lambda_F \subset \partial\Omega$ in Theorem 4.1. Moreover we have case (C) by Corollary 3.18. Then (4.1) becomes

$$\operatorname{div}(\varphi F)(\Omega) = \int_{\partial\Omega} \varphi\, d\sigma_F \tag{4.16}$$

for a Radon measure σ_F supported on $\partial\Omega$.

For the open set Ω we now consider $\varphi \in \mathcal{W}^{1,\infty}(U)$ supported outside $B_r(0)$ for small $r > 0$. Then the usual Gauss-Green formula gives

$$\operatorname{div}(\varphi F)(\Omega) = \int_{\partial\Omega} \varphi F \nu^\Omega\, d\mathcal{H}^1 = \int_{\partial\Omega} \varphi\, d\sigma_F.$$

Since $\mathcal{W}^{1,\infty}(U)$ is dense in $C(U)$, we obtain

$$\sigma_F = F\nu^\Omega \mathcal{H}^1 \lfloor \partial\Omega \quad \text{on} \quad \partial\Omega \setminus \{0\}.$$

From (4.16) we get with $\varphi = 1$ on U

$$0 = \operatorname{div} F(\Omega) = \sigma_F(\{0\}) + \int_{\partial B_1(0)\cap\partial\Omega} F\nu^\Omega\, d\mathcal{H}^1 = \sigma_F(\{0\}) + \frac{\alpha}{2\pi}.$$

Hence

$$\sigma_F(\{0\}) = -\frac{\alpha}{2\pi}$$

and, therefore,

$$\sigma_F = F\nu^\Omega \mathcal{H}^1 \llcorner \partial\Omega - \tfrac{\alpha}{2\pi}\delta_0 \,. \tag{4.17}$$

For $\overline{\Omega}$ instead of Ω we can argue the same way. Then the concentration at the origin is contained. This leads to (4.16) with $\overline{\Omega}$ and some measure $\overline{\sigma}_F$ that is given by

$$\overline{\sigma}_F = F\nu^\Omega \mathcal{H}^1 \llcorner \partial\Omega + \tfrac{2\pi-\alpha}{2\pi}\delta_0 \,.$$

Notice that

$$\sigma_F(\{0\}) = \lim_{r\to 0} \int_{\partial B_r(0)\cap\Omega} F\nu^{B_r(0)^c}\, d\mathcal{H}^1$$

but

$$\overline{\sigma}_F(\{0\}) = \lim_{r\to 0} \int_{\partial B_r(0)\cap\Omega^c} F\nu^{B_r(0)}\, d\mathcal{H}^1 \,.$$

This allows the interpretation that σ_F is a trace from inside, i.e. it can be computed from F restricted to the interior of Ω and this way disregards the concentration of $\operatorname{div} F$ on the boundary, while $\overline{\sigma}_F$ is a trace from outside, i.e. it can be computed from F restricted to the exterior of Ω and includes the concentration of $\operatorname{div} F$ at the origin. However we have that $\sigma_F(\{0\})$ and $\overline{\sigma}_F(\{0\})$ depend on the opening angle α and, in (4.16), they merely contribute a part of the concentration of $\operatorname{div} F$.

We can also treat any Borel set $\tilde{\Omega}$ with

$$\Omega \subset \tilde{\Omega} \subset \overline{\Omega} \,.$$

Then we readily get for the associated measure $\tilde{\sigma}_F$ that

$$\tilde{\sigma}_F = \begin{cases} \sigma_f & \text{if } 0 \notin \tilde{\Omega}\,, \\ \overline{\sigma}_F & \text{if } 0 \in \tilde{\Omega}\,. \end{cases}$$

This exact treatment of concentrations on the boundary allows in applications e.g. a very precise description how a point load at a body is balanced by its parts (cf. [33], [38, Example 2]). Moreover it shows that a dependence of such a concentration on a normal at $x = 0$ does not make sense. Notice that point concentrations can also occur at a cuspidate corner and need not be proportional to the opening angle α in general (cf. [34]).

As limit case $\alpha = 0$ we finally consider

$$\Omega = (0, 1) \times \{0\} \,.$$

Then we have for all $\varphi \in \mathcal{W}^{1,\infty}(U)$

$$\operatorname{div}(\varphi F)(\Omega) = \int_{\Omega} \varphi \, d \operatorname{div} F = 0\,, \quad \operatorname{div}(\varphi F)(\overline{\Omega}) = \varphi(0)\,.$$

Hence we readily get with the zero measure σ_0

$$\operatorname{div}(\varphi F)(\Omega) = \int_{\partial\Omega} \varphi \, d\sigma_0 \quad \text{and} \quad \operatorname{div}(\varphi F)(\overline{\Omega}) = \int_{\partial\Omega} \varphi \, d\delta_0\,.$$

This agrees with σ_F and $\overline{\sigma}_F$ from above if we take as outer "unit" normal of Ω the sum of the upward and the downward normal, i.e.

$$\nu^{\Omega} = \left(\begin{smallmatrix} 0 \\ 1 \end{smallmatrix}\right) + \left(\begin{smallmatrix} 0 \\ -1 \end{smallmatrix}\right) = \left(\begin{smallmatrix} 0 \\ 0 \end{smallmatrix}\right).$$

Obviously we can also treat $\Omega = \{0\}$ that way.

In the light of $\tilde{\Omega}$ in the previous example let us provide a further simple example demonstrating how the Gauss-Green formula exactly takes care of the points belonging to Ω.

Example 4.13 For $U = B_2(0) \subset \mathbb{R}^2$ we consider $F \in \mathcal{DM}^1(U)$ given by

$$F = \begin{cases} (2,0) & \text{on } (0,1)^2\,, \\ (1,0) & \text{otherwise}\,. \end{cases}$$

The first component of F is a BV function and we readily get

$$\operatorname{div} F = \mathcal{H}^1 \llcorner \{0\} \times (0,1) - \mathcal{H}^1 \llcorner \{1\} \times (0,1)$$

(cf. [22, p. 169]). Let Ω be a Borel set satisfying

$$(0,1)^2 \subset \Omega \subset [0,1]^2\,.$$

By Corollary 3.18 and Proposition 4.7 we have case (C) for small $\delta > 0$ and we can choose (4.1) in the form

$$\operatorname{div}(\varphi F)(\Omega) = \int_{\partial\Omega} \varphi \, d\sigma_F \quad \text{for all} \quad \varphi \in \mathcal{W}^{1,\infty}(U)$$

with a Radon measure σ_F supported on $\partial\Omega$. For the determination of σ_F we consider rectangles $R \subset U$ intersecting $\partial\Omega$. Then we approximate χ_R by

$$\chi_R^{\delta} := \chi_R + \chi_{R_\delta \setminus R}\big(1 - \tfrac{1}{\delta} \operatorname{dist}_R\big) \in \mathcal{W}^{1,\infty}(U) \;\text{ for small }\; \delta > 0$$

and use $\operatorname{div}(\chi_R^\delta F) = \chi_R^\delta \operatorname{div} F + F \cdot D\chi_R^\delta$. This way we get with

$$\Gamma_\Omega := \partial\Omega \cap \Omega\,, \quad F_{\text{int}} := (2,0)\,, \quad F_{\text{ext}} := (1,0)\,,$$

that

$$\operatorname{div}(\varphi F)(\Omega) = \int_{\partial\Omega\setminus\Gamma_\Omega} \varphi F_{\text{int}} \cdot \nu^\Omega \, d\mathcal{H}^1 + \int_{\Gamma_\Omega} \varphi (F_{\text{int}} - F_{\text{ext}}) \cdot \nu^\Omega \, d\mathcal{H}^1$$

for all $\varphi \in \mathcal{W}^{1,\infty}(U)$.

In Example 4.10 we have already seen that the choice of λ_F and μ_F in (4.1) is not unique in general. For a simple case we now demonstrate that the usual boundary integral $\int_{\partial\Omega} \varphi F \cdot \nu \, d\mathcal{H}^{n-1}$ in the Gauss-Green formula can be completely replaced by $\int_{\partial\Omega} D\varphi \, d\mu_F$ for some suitable μ_F.

Example 4.14 For $U = \Omega = (-1,1)^2 \subset \mathbb{R}^2$ and the vector field

$$F(x,y) = \begin{pmatrix} |x| \\ 0 \end{pmatrix} \quad \text{on} \quad \Omega$$

we have classically

$$\int_\Omega F D\varphi \, d\mathcal{L}^n + \int_\Omega \varphi \operatorname{div} F \, d\mathcal{L}^n = \int_{\partial\Omega} \varphi F \cdot \nu^\Omega \, d\mathcal{H}^1 \tag{4.18}$$

for all $\varphi \in \mathcal{W}^{1,\infty}(\Omega)$. For a transformation of the right hand side we first consider $\varphi \in C^1(\overline{\Omega})$ and use integration by parts piecewise on $\partial\Omega$ to get

$$\begin{aligned}
\int_{\partial\Omega} \varphi F \cdot \nu^\Omega \, d\mathcal{H}^1 &= \int_{-1}^1 \varphi(1,y) - \varphi(-1,y) \, dy \\
&= -\int_{-1}^1 y\varphi_y(1,y) - y\varphi_y(-1,y) \, dy \\
&\quad + \big[y\varphi(1,y) - y\varphi(-1,y) \big]_{-1}^1 \\
&= -\int_{-1}^1 y\varphi_y(1,y) - y\varphi_y(-1,y) \, dy \\
&\quad + \int_{-1}^1 \varphi_x(x,1) + \varphi_x(x,-1) \, dx \\
&= \int_{\partial\Omega} G D\varphi \, d\mathcal{H}^1
\end{aligned} \tag{4.19}$$

where

$$G(x,y) = \begin{cases} (0, -xy) & \text{for } |x| > |y|\,, \\ (|y|, 0) & \text{for } |x| < |y|\,. \end{cases}$$

For some general $\varphi \in \mathcal{W}^{1,\infty}(\Omega)$ we choose smooth $\varphi_k \in C^\infty(\overline{\Omega})$ approximating φ within $\mathcal{W}^{1,1}(\Omega)$ and $\mu_{\partial\Omega}$ be the measure from Proposition 3.1 related to $E = \Omega$ and $\gamma(\delta) = \delta$. Piecewise continuity of the integrand up to the boundary, the coarea formula, and (3.6) imply

$$\int_{\partial\Omega} GD\varphi_k \, d\mathcal{H}^1 = \lim_{\delta\downarrow 0} \frac{1}{\delta} \int_{(\partial\Omega)_\delta \cap \Omega} GD\varphi_k \, d\mathcal{L}^2 = \fint_{\partial\Omega} GD\varphi_k \, d\mu_{\partial\Omega} \quad \text{for all } k\,.$$

One readily checks for small $\delta > 0$ that (4.19) is also valid with $\Omega_{-\delta}$ instead of Ω. Moreover we extend ν^Ω onto Ω by setting $\nu^\Omega = \nu^{\Omega_{-\delta}}$ on $\partial\Omega_{-\delta}$. Then the coarea formula yields

$$\frac{1}{\delta} \int_{(\partial\Omega)_\delta \cap \Omega} \varphi_k F \cdot \nu^\Omega \, d\mathcal{L}^2 = \frac{1}{\delta} \int_{(\partial\Omega)_\delta \cap \Omega} GD\varphi_k \, d\mathcal{L}^2 \quad \text{for all } k\,.$$

By the uniform convergence $\varphi_k \to \varphi$, the limit $k \to \infty$ on the left hand side is uniform in δ. With $D\varphi_k \to D\varphi$ in $\mathcal{L}^1(\Omega)$ we therefore get that the limit

$$\lim_{k\to\infty} \frac{1}{\delta} \int_{(\partial\Omega)_\delta \cap \Omega} GD\varphi_k \, d\mathcal{L}^2 = \frac{1}{\delta} \int_{(\partial\Omega)_\delta \cap \Omega} GD\varphi \, d\mathcal{L}^2$$

is also uniform in δ. Using Corollary 3.2 we get

$$\begin{aligned} \int_{\partial\Omega} \varphi F \cdot \nu^\Omega \, d\mathcal{H}^1 &= \lim_{k\to\infty} \int_{\partial\Omega} \varphi_k F \cdot \nu^\Omega \, d\mathcal{H}^1 = \lim_{k\to\infty} \int_{\partial\Omega} GD\varphi_k \, d\mathcal{H}^1 \\ &= \lim_{k\to\infty} \fint_{\partial\Omega} GD\varphi_k \, d\mu_{\partial\Omega} = \fint_{\partial\Omega} GD\varphi \, d\mu_{\partial\Omega}\,. \end{aligned}$$

Thus we end up with

$$\int_\Omega FD\varphi \, d\mathcal{L}^n + \int_\Omega \varphi \operatorname{div} F \, d\mathcal{L}^n = \fint_{\partial\Omega} GD\varphi \, d\mu_{\partial\Omega} \tag{4.20}$$

for all $\varphi \in \mathcal{W}^{1,\infty}(\Omega)$. From (4.19) we see that we can replace $\mu_{\partial\Omega}$ with $\mathcal{H}^1 \llcorner \partial\Omega$ for $\varphi \in C^1(\overline{\Omega})$. But notice that $\mu_{\partial\Omega}$ is pure by Proposition 2.12 and that we cannot take a σ-measure in (4.20) in general. Finally it is a simple observation by taking $\varphi = 1$ on Ω that a Gauss-Green formula like (4.20) is only possible if $(\operatorname{div} F)(\Omega) = 0$ (while $\operatorname{div} F = 0$ is not necessary).

Let us also give some more general example in $\mathbb{R}^3$ for a Gauss-Green formula of the form (4.20). We refrain from looking for the most general version, since we just want to provide the essential idea for a class of vector fields satisfying $\operatorname{div} F = 0$.

Example 4.15 Let $U \subset \mathbb{R}^3$ be open and bounded, let $\Omega \Subset U$ be open and connected with smooth boundary, and let $F \in C^1(U, \mathbb{R}^3)$ be a bounded vector field such that

$$F = \operatorname{curl} G \quad \text{where} \quad G \in C^2(U, \mathbb{R}^3)\,.$$

Obviously $\operatorname{div} F = 0$ and thus, $F \in \mathcal{DM}^1(U)$. By the smooth boundary we have (4.11). Hence, by Proposition 4.7, we can choose $\mu_F = 0$ and λ_F with core $\lambda_F \subset \partial\Omega$ in Theorem 4.1 while the usual Gauss-Green formula gives

$$\operatorname{div}(\varphi F)(\Omega) = \int_\Omega F \cdot D\varphi \, d\mathcal{L}^3 = \int_{\partial\Omega} \varphi F \cdot \nu^\Omega \, d\mathcal{H}^2 \tag{4.21}$$

for all $\varphi \in C^1(U)$. Now we find two disjoint smooth manifolds $\Gamma_j \subset \partial\Omega$ with smooth boundary such that $\partial\Omega = \overline{\Gamma}_1 \cup \overline{\Gamma}_2$ (since $\partial\Omega$ is locally the graph of a smooth function, we can intersect $\partial\Omega$ transversally with a small circular cylinder to get the desired decomposition into a 'big' and a 'small' manifold). By the classical Stoke's theorem and a coherent orientation of the boundaries $\partial\Gamma_j$ with some tangent field t_j we have

$$\int_{\Gamma_j} \operatorname{curl}(\varphi G) \cdot \nu^\Omega \, d\mathcal{H}^2 = \int_{\partial\Gamma_j} \varphi G \cdot t_j \, d\mathcal{H}^1$$

for all $\varphi \in C^1(U)$. From

$$\operatorname{curl}(\varphi G) = \varphi \operatorname{curl} G + D\varphi \times G$$

we obtain some kind of integration by parts on the manifolds Γ_j

$$\int_{\Gamma_j} (\varphi \operatorname{curl} G) \cdot \nu^\Omega \, d\mathcal{H}^2 + \int_{\Gamma_j} (D\varphi \times G) \cdot \nu^\Omega \, d\mathcal{H}^2 = \int_{\partial\Gamma_j} \varphi G \cdot t_j \, d\mathcal{H}^1\,.$$

If we sum up the integrals on Γ_1 and Γ_2, the boundary terms cancel out by $t_1 = -t_2$. Moreover,

$$(D\varphi \times G) \cdot \nu^\Omega = (G \times \nu^\Omega) \cdot D\varphi\,.$$

Thus, (4.21) becomes

$$\operatorname{div}(\varphi F)(\Omega) = \int_\Omega F \cdot D\varphi \, d\mathcal{L}^3 = -\int_{\partial\Omega} (G \times \nu^\Omega) \cdot D\varphi \, d\mathcal{H}^2$$

for all $\varphi \in C^1(U)$. For $\varphi \in \mathcal{W}^{1,\infty}(U)$ we consider $\mu_{\partial\Omega}$ from Proposition 3.1 with $E = \Omega$ and $\gamma(\delta) = \delta$. By arguments as in Example 4.14 we can apply Corollary 3.2 to get

$$\operatorname{div}(\varphi F)(\Omega) = \int_{\Omega} F \cdot D\varphi \, d\mathcal{L}^3 = -\!\!\!\!\!\!\int_{\partial\Omega} \left(G \times \nu^{\Omega}\right) \cdot D\varphi \, d\mu_{\partial\Omega}$$

for all $\varphi \in \mathcal{W}^{1,\infty}(U)$.

In contrast to Example 4.10 we now consider a constant vector field F and a similar open set $\Omega \subset \mathbb{R}^2$ having infinite perimeter but such that (4.5) is satisfied. This tells us that the choice of the measures λ_F, μ_F independent of $\delta > 0$ is not restricted to Ω with finite perimeter.

Example 4.16 In $\mathbb{R}^2$ we consider the open set

$$U = \Omega := \bigcup_{k=1}^{\infty} R_k \quad \text{with} \quad R_k := \left(\tfrac{1}{k}, \tfrac{1}{k} + \tfrac{1}{k^2}\right) \times \left(0, \tfrac{1}{k}\right)$$

and the constant vector field

$$F \in \mathcal{L}^1(U, \mathbb{R}^2) \quad \text{with} \quad F = \left(\begin{smallmatrix} 1 \\ 0 \end{smallmatrix}\right).$$

Since $\frac{1}{k} + \frac{1}{k^2} < \frac{1}{k-1}$, the R_k are pairwise disjoint. Clearly $F \in \mathcal{DM}^1(U)$ and Ω has infinite perimeter. In order to verify (4.9) we first study the supremum in (4.9) over

$$M_\delta := \left\{\varphi \in \mathcal{W}^{1,\infty}(U) \;\middle|\; \|\varphi_{|(\partial\Omega)_\delta \cap \Omega}\|_{\mathcal{W}^{1,\infty}} \le 1\right\}$$

merely on some fixed R_k for some given $\delta > 0$. The coarea formula gives that

$$\int_{(\partial R_k)_\delta \cap R_k} \varphi F D\chi_\delta^{\partial\Omega} \, d\mathcal{L}^2 = \int_{\frac{\delta}{2}}^{\delta} \int_{\partial((R_k)_{-\tau})} \varphi F D\chi_\delta^{\partial\Omega} \, d\mathcal{H}^1 \, d\tau \tag{4.22}$$

where the inner integral on the right hand side vanishes if $(R_k)_{-\tau}$ is empty. For "relevant" τ that inner integral has to be taken on the boundary of a (translated) rectangle of the form

$$R := (-a, a) \times (0, b) \quad \text{with} \quad 0 < a < \tfrac{1}{2k^2}, \; 0 < b < \tfrac{1}{k}. \tag{4.23}$$

The integral vanishes on the short sides of R by $F D\chi_\delta^{\partial\Omega} = 0$ and one has

$$F D\chi_\delta^{\partial\Omega} = \pm\tfrac{2}{\delta} \quad \text{on} \quad \Gamma_\pm := \{\pm a\} \times (0, b).$$

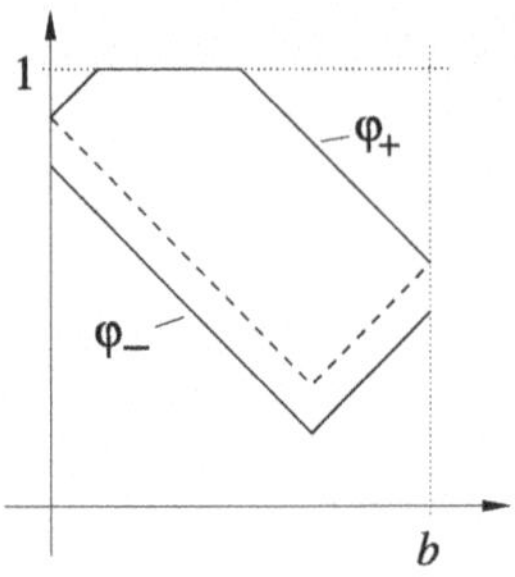

Fig. 4.1 The figure shows the graphs of φ_+ and φ_- where the dashed graph is a translation of φ_-

For an estimate of the supremum in (4.9) we consider $\mathcal{W}^{1,\infty}$-functions φ on a small neighborhood of ∂R with $\|\varphi\|_\infty \leq 1$ and $\|D\varphi\|_\infty \leq 1$. Using $\varphi_\pm(y) := \varphi(\pm a, y)$ we want to maximize

$$\int_0^b \varphi_+ - \varphi_- \, dy \tag{4.24}$$

for such φ (cf. Fig. 4.1). Thus we have to look for φ with $\varphi_+ \geq \varphi_-$ such that the area between the graphs of $\varphi_\pm$ becomes maximal. Since an additive constant for φ does not change the integral, we can assume that the maximum of φ_+ on $[0, b]$ equals 1. Hence, by $\|D\varphi\|_\infty \leq 1$ and the size of R, we always have $\varphi \geq -1$, i.e. we do not have to take care explicitly for that constraint. Let us now briefly assume that $\varphi_+(0)$, $\varphi_+(b)$ are fixed. For such φ_+ we denote the smallest $y \in [0, b]$ with $\varphi_+(y) = 1$ by y_1 and the largest by y_2. In order to maximize the integral for such φ_+, we use

$$\varphi_+(y) = \varphi_+(0) + \int_0^y \varphi'_+(\tau)\, d\tau$$

to see that φ'_+ has to equal 1 on $[0, y_1]$. Analogously φ'_+ has to equal -1 on $[y_2, b]$ and, clearly, $\varphi_+ = 1$ on $[y_1, y_2]$. The same way we get that the optimal φ_- with fixed values on the boundary first has to decay with slope -1 and then it grows with slope 1, where we have used that always $\varphi_- \geq -1$ (cf. Fig. 4.1). If φ_+ equals 1 on a nontrivial interval, then we can enlarge the integral in (4.24) by a translation of φ such that $\varphi_+ = 1$ merely at a single point $\tilde{y} \in [0, b]$. Moreover, for a maximal value in (4.24) the φ should have maximal slope on the short sides of R such that

$$\varphi_-(0) = \varphi_+(0) - 2a\,, \quad \varphi_-(b) = \varphi_+(b) - 2a\,.$$

Now it is sufficient to look for an optimal φ in this subclass of admissible φ. Clearly the "shape" of the optimal $\varphi_\pm$ is not influenced if we add $2a$ to φ_- for a moment (cf. dashed graph in Fig. 4.1) such that the graphs of $\varphi_\pm$ form a rectangle with sides having length $\sqrt{2}\tilde{y}$ and $\sqrt{2}(b - \tilde{y})$. Hence the sum is independent of $\tilde{y}$ and always equals $\sqrt{2}b$. Therefore the area of the rectangle between the graphs becomes maximal if it is a square. Consequently there is some φ that maximizes (4.24) under the considered constraint and it satisfies (with the actual not shifted graph of φ_-)

$$\varphi_+\left(\tfrac{b}{2}\right) = 1\,,\quad \varphi_+(0) = \varphi_+(b) = 1 - \tfrac{b}{2}\,,$$

$$\varphi_-(0) = \varphi_-(b) = 1 - \tfrac{b}{2} - 2a\,,\quad \varphi_-\left(\tfrac{b}{2}\right) = 1 - b - 2a\,.$$

Therefore

$$\int_0^b \varphi_+ - \varphi_-\,dy = \left(\frac{\sqrt{2}b}{2}\right)^2 + 2ab = \frac{b^2}{2} + 2ab$$

for the maximal φ. Since $(R_k)_{-\tau} \subset R_k$ is a nonempty rectangle for $\tau < \frac{1}{2k^2}$, we can use the bounds from (4.23) to get

$$\sup_{\varphi\in M_\delta} \int_{\partial((R_k)_{-\tau})} \varphi F D\chi_\delta^{\partial\Omega}\,d\mathcal{H}^1 \le \sup_{\varphi\in M_\delta} \int_{\partial R_k} \varphi F D\chi_\delta^{\partial\Omega}\,d\mathcal{H}^1 = \frac{1}{\delta k^2} + \frac{2}{\delta k^3}$$

for $\tau \in \left[\frac{\delta}{2}, \delta\right] \cap \left(0, \frac{1}{2k^2}\right)$. Since the left hand side vanishes if $(R_k)_{-\tau} = \emptyset$, the estimate is even true for a.e. $\tau \in \left[\frac{\delta}{2}, \delta\right]$. Though a "good" $\varphi \in M_\delta$ for the supremum in (4.9) might not be optimal for each $\partial((R_k)_{-\tau})$, from (4.22) we get at least the estimate

$$\sup_{\varphi\in M_\delta} \int_{(\partial R_k)_\delta \cap R_k} \varphi F D\chi_\delta^{\partial\Omega}\,d\mathcal{L}^n \le \frac{1}{2k^2} + \frac{1}{k^3} \quad \text{for all} \quad k \in \mathbb{N}\,,\ \delta > 0\,.$$

Hence

$$\sup_{\varphi\in M_\delta} \int_{(\partial\Omega)_\delta \cap \Omega} \varphi F D\chi_\delta^{\partial\Omega}\,d\mathcal{L}^n \le \sum_{k=1}^\infty \left(\frac{1}{2k^2} + \frac{1}{k^3}\right) < \infty \quad \text{for all} \quad \delta > 0\,.$$

But this implies (4.9). Thus, according to Proposition 4.5 and Lemma 4.6, we can choose λ_F, μ_F in (4.1) independent of δ with core in $\partial\Omega$.

Assume that $\mu_F = 0$. Then we can consider functions $\varphi \in \mathcal{W}^{1,\infty}(\Omega)$ that vanish outside some fixed R_k. These φ are continuously extendable onto $\overline{R_k}$. From the usual Gauss-Green formula on R_k we obtain that the restriction of λ_F on R_k is related to the Radon measure

$$\sigma_k = F\nu^{R_k} \llcorner \partial R_k \quad \text{where} \quad |\lambda_F|(R_k) = |\sigma_k|(\partial R_k) = \tfrac{2}{k}$$

(cf. Proposition 2.13). Taking the sum, this is a contradiction, since $|\lambda_F|(\Omega)$ would not be finite. Therefore $\mu_F \neq 0$. Let us briefly sketch how the measures λ_F and μ_F

can be chosen. We fix some $\tilde{k} \in \mathbb{N}$ and set

$$\Omega_1 := \bigcup_{k=1}^{\tilde{k}} R_k\,, \quad \Omega_2 := \bigcup_{k=\tilde{k}+1}^{\infty} R_k\,.$$

Obviously Ω_1 is an open set with Lipschitz boundary and we can choose λ_F corresponding to the Radon measure σ_F on $\partial\Omega_1$ from the boundary integral in the usual Gauss-Green formula on Ω_1 (cf. also Example 4.10). For μ_F we first argue similar to Example 4.14 on some fixed R_k to get for $\varphi \in C^1(\overline{R_k})$

$$\begin{aligned}\int_{\partial R_k} \varphi F \cdot \nu^{R_k}\, d\mathcal{H}^1 &= \int_0^{\frac{1}{k}} \varphi(\tfrac{1}{k}+\tfrac{1}{k^2}, y) - \varphi(\tfrac{1}{k}, y)\, dy \\ &= \int_0^{\frac{1}{k}} y\big(\varphi_y(\tfrac{1}{k}, y) - \varphi_y(\tfrac{1}{k}+\tfrac{1}{k^2}, y)\big)\, dy \\ &\quad + \tfrac{1}{k}\big(\varphi(\tfrac{1}{k}+\tfrac{1}{k^2}, \tfrac{1}{k}) - \varphi(\tfrac{1}{k}, \tfrac{1}{k})\big) \\ &= \int_0^{\frac{1}{k}} y\big(\varphi_y(\tfrac{1}{k}, y) - \varphi_y(\tfrac{1}{k}+\tfrac{1}{k^2}, y)\big)\, dy \\ &\quad + \tfrac{1}{k}\int_{\frac{1}{k}}^{\frac{1}{k}+\frac{1}{k^2}} \varphi_x(x, \tfrac{1}{k})\, dx \\ &= \int_{\partial R_k} G D\varphi\, d\mathcal{H}^1\end{aligned}$$

with a suitable vector field G on $\overline{R_k}$. The measure

$$G\mathcal{H}^1 \llcorner \partial R_k \quad \text{has total variation} \quad \tfrac{1}{k^2} + \tfrac{1}{k^3}$$

and we can construct μ_F on R_k similar to Example 4.14. Summing up over $k > \tilde{k}$ we finally get μ_F on Ω_2.

Let us come back to Example 4.9. We construct an uncountable family of measures $(\lambda_F^\rho, \mu_F^\rho)$ satisfying (4.1) where even $\lambda_F = 0$ is possible.

Example 4.17 Let $U = B_2(0)$ as in Example 4.9, but for technical simplicity we now consider the cone

$$\Omega := (0,\infty)^2 \cap B_1(0) \subset \mathbb{R}^2\,.$$

For the vector field

$$F(x,y) := \frac{1}{2\pi(x^2+y^2)}\begin{pmatrix}-y\\ x\end{pmatrix}$$

we already now that $F \in \mathcal{DM}^1(U)$ with $\operatorname{div} F = 0$ and that $\mu_F \neq 0$ is impossible in (4.1). Moreover we have (4.5) and case (C) for small $\delta > 0$. For any $\rho \in (0,1]$ we now construct Radon measures σ_F^ρ supported on $\partial\Omega$ and μ_F^ρ in $\operatorname{ba}(U, \mathcal{B}(U), \mathcal{L}^2)^2$ with core in $\partial\Omega$ such that

$$\operatorname{div}(\varphi F)(\Omega) = \int_{\partial\Omega} \varphi \, d\sigma_F^\rho + \fint_{\partial\Omega} D\varphi \, d\mu_F^\rho \quad \text{for all} \quad \varphi \in \mathcal{W}^{1,\infty}(U)\,.$$

Notice that the measures μ_F^ρ are pure by Proposition 2.12.

For fixed $\rho \in (0,1]$ we now set

$$\Omega^\delta := \big(\Omega \setminus B_\delta(0)\big) \cap B_\rho(0), \quad \tilde{\Omega} := \Omega \setminus \overline{B_\rho(0)} \quad \text{for} \quad \delta \in (0,\rho)\,.$$

Obviously F is tangential to circles around the origin. Then, by the classical Gauss-Green theorem, we get for all $\varphi \in \mathcal{W}^{1,\infty}(U)$

$$\begin{aligned} \operatorname{div}(\varphi F)(\Omega^\delta \cup \tilde{\Omega}) &= \int_{\partial\Omega^\delta} \varphi F \nu^{\Omega^\delta} \, d\mathcal{H}^1 + \int_{\partial\tilde{\Omega}} \varphi F \nu^{\tilde{\Omega}} \, d\mathcal{H}^1 \\ &= I_\delta + \int_{\partial\tilde{\Omega}} \varphi \, d\tilde{\sigma}_F \overset{\delta\to 0}{\longrightarrow} \operatorname{div}(\varphi F)(\Omega) \end{aligned}$$

where $\tilde{\sigma}_F$ is the Radon measure

$$\begin{aligned} \tilde{\sigma}_F &= -\frac{1}{2\pi x}\mathcal{H}^1 \llcorner \big([\rho,1] \times \{0\}\big) + \frac{1}{2\pi y}\mathcal{H}^1 \llcorner \big(\{0\} \times [\rho,1]\big) \qquad \text{and} \\ I_\delta &= \frac{1}{2\pi}\bigg(\int_\delta^\rho \frac{\varphi(0,0) - \varphi(x,0)}{x}\,dx + \int_\delta^\rho \frac{\varphi(0,y) - \varphi(0,0)}{y}\,dy\bigg) \qquad (4.25) \end{aligned}$$

where $\varphi(0,0)$ is included additionally. We use the Lipschitz continuous function

$$\psi(x,y) := \varphi(0,y) - \varphi(x,y) \quad \text{satisfying} \quad |\psi(x,y)| \le x\,\|D\varphi\|_\infty$$

to treat the first integral in I_δ. Though the limit of the integral for $\delta \to 0$ exists by dominated convergence, the limit of the measures $\frac{1}{x}\mathcal{L}^1 \llcorner [\delta,\rho]$ for $\delta \to 0$ does not give a finite measure. Hence the limit of I_δ cannot contribute to σ_F. We therefore use integration by parts to get a contribution to μ_F. The difficulty that $D\varphi$ might not exist on $\partial\Omega^\delta$ is circumvented by fattening up the boundary. Clearly, $\psi(\cdot, y)$ is absolutely continuous for all y. Since $D\varphi$ exists $\mathcal{L}^2$-a.e., the fundamental theorem of calculus implies

$$\psi(x,y)\ln x\Big|_{x=\delta}^{\rho} = \int_\delta^\rho \Big(\psi_x(x,y)\ln x + \psi(x,y)\frac{1}{x}\Big)\,dx \qquad \text{for a.e. } y \tag{4.26}$$

(cf. [53, p. 1019], [22, pp. 164, 235]). Notice that

$$|\psi(x,y)\ln x| \le x|\ln x|\,\|D\varphi\|_\infty \overset{x\to 0}{\longrightarrow} 0\,, \qquad \Big|\frac{\psi(x,y)}{x}\Big| \le \|D\varphi\|_\infty\,.$$

Using dominated convergence and $\psi_x = -\varphi_x$, we can take the limit $\delta \to 0$ in (4.26). Then we integrate over $[0,\tau]$ with respect to y and divide everything by τ to get

$$\frac{1}{\tau}\int_0^\tau \psi(\rho,y)\ln\rho\,dy = \int_{[0,\rho]\times[0,\tau]} D\varphi\,d\mu'_{F,\tau} + \frac{1}{\tau}\int_0^\tau\int_0^\rho \psi(x,y)\frac{1}{x}\,dxdy \tag{4.27}$$

with the vector measures

$$\mu'_{F,\tau} = -\tfrac{1}{\tau}\big(\begin{smallmatrix}\ln x\\ 0\end{smallmatrix}\big)\mathcal{L}^2\llcorner\big([0,\rho]\times[0,\tau]\big) \in \mathrm{ba}(U,\mathcal{B}(U),\mathcal{L}^2)^2\,.$$

Obviously

$$|\mu'_{F,\tau}|(U) \le \int_0^1 |\ln x|\,dx = 1 \quad \text{for all } \tau > 0\,, \quad \mathrm{core}\,\mu'_{F,\tau} \subset [0,\rho]\times[0,\tau]\,.$$

Hence, by the Alaoglu theorem, the measures $\{\mu'_{F,\frac{1}{k}}\}_k$ have a weak* cluster point

$$\mu'_F \in \mathrm{ba}(U,\mathcal{B}(U),\mathcal{L}^2)^2 \quad \text{with} \quad \mathrm{core}\,\mu'_F \subset \partial\Omega \cap B_\rho(0)\,.$$

In (4.27) the limit for $\tau \to 0$ exists for the most left and the most right integral by continuity of ψ (notice that $\frac{\psi(x,y)}{x}$ is bounded and continuous on $(0,\rho]\times[0,\tilde\tau]$ for some $\tilde\tau > 0$ and that $\frac{1}{\tau}\int_0^\tau \frac{\psi(x,y)}{x}\,dy \to \frac{\psi(x,0)}{x}$ as $\tau \to 0$ for all $x \in (0,\rho)$). Hence we can take the limit $\tau \to 0$ in (4.27) and, with the definition of ψ, we obtain

$$\big(\varphi(0,0)-\varphi(\rho,0)\big)\ln\rho = \int\!\!\!\!\!\!-_{[0,\rho]\times\{0\}} D\varphi\,d\mu'_F + \int_0^\rho \frac{\varphi(0,0)-\varphi(x,0)}{x}\,dx \tag{4.28}$$

for all $\varphi \in \mathcal{W}^{1,\infty}(U)$. Analogous arguments for the second integral in (4.25) give a weak* cluster point μ''_F of the vector measures

$$\mu''_{F,\tau} = \tfrac{1}{\tau}\big(\begin{smallmatrix}0\\ \ln y\end{smallmatrix}\big)\mathcal{L}^2\llcorner\big([0,\tau]\times[0,\rho]\big) \in \mathrm{ba}(U,\mathcal{B}(U),\mathcal{L}^2)^2$$

with core $\mu''_F \subset \partial\Omega \cap B_\rho(0)$. Now we use (4.28) and the analogous equation with μ''_F to replace the two integrals in (4.25). Then, with

$$\mu_F^\rho := -\tfrac{1}{2\pi}(\mu'_F + \mu''_F), \quad \sigma_F^\rho := \tilde{\sigma}_F + \tfrac{\ln\rho}{2\pi}\big(\delta_{(0,\rho)} - \delta_{(\rho,0)}\big),$$

we finally obtain

$$\begin{aligned} \operatorname{div}(\varphi F)(\Omega) &= \lim_{\delta\to 0} I_\delta + \int_{\partial\tilde{\Omega}} \varphi \, d\tilde{\sigma}_F \\ &= \int_{\partial\Omega} \varphi \, d\sigma_F^\rho + \fint_{\partial\Omega\cap B_\rho(0)} D\varphi \, d\mu_F^\rho \end{aligned} \tag{4.29}$$

for all $\varphi \in \mathcal{W}^{1,\infty}(U)$ and all $\rho \in (0, 1]$, where

$$\int_{\partial\Omega} \varphi \, d\sigma_F^\rho = \int_{\partial\Omega\setminus B_\rho(0)} \varphi F \nu^\Omega \, d\mathcal{H}^1 + \frac{\ln\rho}{2\pi}\big(\varphi(0,\rho) - \varphi(\rho,0)\big).$$

Notice that this covers the special case $\sigma_F^1 = 0$ for $\rho = 1$. For $\varphi \in C^1(U)$ we can argue by continuity to get

$$\fint_{\partial\Omega\cap B_\rho(0)} D\varphi \, d\mu_F^\rho = \frac{1}{2\pi}\Big(\int_0^\rho \varphi_x(x,0)\ln x \, dx - \int_0^\rho \varphi_y(0,y)\ln y \, dy\Big).$$

Moreover, from (4.29) with $\varphi = 1$ on U, we get the Gaussian formula

$$0 = \int_\Omega \operatorname{div} F \, d\mathcal{L}^n = \int_{\partial\Omega\setminus B_\rho(0)} \varphi F \nu^\Omega \, d\mathcal{H}^1$$

for any $\rho \in (0, 1]$. For $\rho = 1$ this includes the exotic special case

$$0 = \int_\Omega \operatorname{div} F \, d\mathcal{L}^n = \int_\emptyset \varphi F \nu^\Omega \, d\mathcal{H}^1 .$$

Let us finally provide an explicit computation for (4.29) with the simple function $\varphi(x, y) = x$ as demonstration. Using polar coordinates for the first volume integral we readily get

$$
\begin{aligned}
-\frac{1}{2\pi} = \int_\Omega F D\varphi \, d\mathcal{L}^2 &= \operatorname{div}(\varphi F)(\Omega) \\
&= -\int_\rho^1 \frac{\varphi(x,0)}{2\pi x}\,dx - \frac{\ln\rho}{2\pi}\varphi(\rho,0) + \int_0^\rho \frac{\varphi_x(x,0)\ln x}{2\pi}\,dx \\
&= -\frac{1-\rho}{2\pi} - \frac{\rho\ln\rho}{2\pi} + \frac{\rho\ln\rho - \rho}{2\pi} = -\frac{1}{2\pi}
\end{aligned}
$$

for all $\rho \in (0,1)$.

Proof of Proposition 4.8 For (1) let $\varphi \in \mathcal{W}^{1,\infty}(U)$ and $\delta \in (0,\tilde{\delta})$ such that (4.14) is satisfied. Hence $\varphi_{|\partial\Omega}$ is Lipschitz continuous on $\partial\Omega$ and there is a Lipschitz continuous extension $\tilde{\varphi}$ onto $\mathbb{R}^n$ with

$$\|\tilde{\varphi}\|_{\mathrm{Lip}(U)} \le \|\varphi_{|\partial\Omega}\|_{\mathrm{Lip}(\partial\Omega)}$$

(cf. [43, p. 452]). By $\tilde{\varphi} = \varphi$ on $\partial\Omega$, Corollary 3.5 implies

$$\operatorname{div}(\varphi F)(\Omega) = \operatorname{div}(\tilde{\varphi} F)(\Omega)\,.$$

Therefore, by (3.17) and (3.32) we get

$$
\begin{aligned}
\operatorname{div}(\chi_\delta^{\partial\Omega}\varphi F)(\Omega) &= \operatorname{div}(\varphi F)(\Omega) = \operatorname{div}(\tilde{\varphi} F)(\Omega) \\
&\le \|F\|_{\mathcal{DM}^1(U)} \|\tilde{\varphi}\|_{\mathcal{W}^{1,\infty}(U)} \\
&\le \|F\|_{\mathcal{DM}^1(U)} \|\tilde{\varphi}\|_{\mathrm{Lip}(U)} \le \|F\|_{\mathcal{DM}^1(U)} \|\varphi_{|\partial\Omega}\|_{\mathrm{Lip}(\partial\Omega)} \\
&\le c\|F\|_{\mathcal{DM}^1(U)} \|\varphi\|_{\mathcal{W}^{1,\infty}((\partial\Omega)_\delta \cap U)}\,.
\end{aligned}
$$

But this readily implies (4.5).

For (2) we verify the assumptions of (1). Obviously $\mathcal{L}^n(\Omega \setminus \operatorname{int}\Omega) = 0$ and all $\varphi \in \mathcal{W}^{1,\infty}(U)$ have a continuous extension onto $\overline{\Omega}$. By definition of Lipschitz boundary we can cover $\partial\Omega$ by finitely many open cylinders C_j with $j = 1, \ldots, m$ such that for Lipschitz continuous functions $\gamma_j : \mathbb{R}^{n-1} \to \mathbb{R}$ and suitable $r_j, h_j > 0$ up to translation and rotation

$$C_j = \{(x',t) \mid |x'| < r_j,\ |t| < h_j\}, \quad |\gamma_j(x')| < \tfrac{h_j}{3},$$

$$\Omega \cap C_j = \{(x',t) \mid |x'| < r_j,\ \gamma_j(x') < t\}$$

where $(x',t) \in \mathbb{R}^{n-1} \times \mathbb{R}$ (cf. [22, pp. 127, 177]). By compactness of $\partial\Omega$ we can find some $\rho > 0$ such that $\rho < \frac{h_j}{3}$ for all j and such that for all $x \in \partial\Omega$ there is j_x

Fig. 4.2 The figure shows the boundary $\partial\Omega$ (solid curve), its neighborhood $(\partial\Omega)_\delta$ (dashed curves), the segment $[x, y]$ in C_{j_x}, and the polygonal curve P in $\overline{\Omega}$

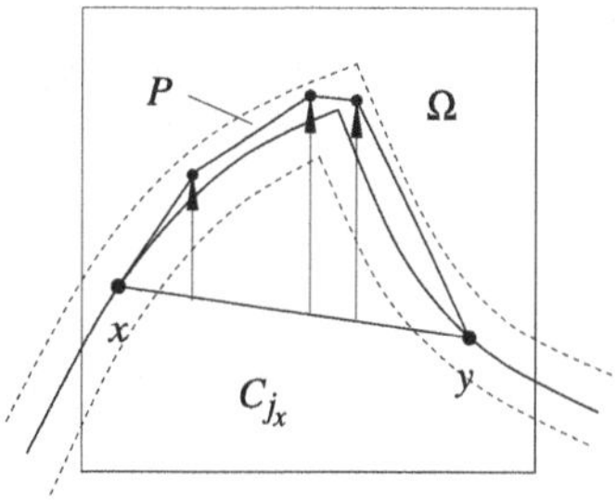

with

$$B_\rho(x) \subset C_{j_x}, \quad j_x \in \{1, \ldots, m\}.$$

Moreover we can assume that $|x - y| > \rho$ if x, y belong to different components of $\overline{\Omega}$. Since all C_j are bounded, there is some $c_0 > 0$ such that

$$\operatorname{diam}(C_j) < c_0\rho \quad \text{for all } j \tag{4.30}$$

where $\operatorname{diam}(C_j)$ denotes the diameter.

Let us now fix $\delta \in (0, \rho)$ and $\varphi \in \mathcal{W}^{1,\infty}(U)$. Then

$$L = \|\varphi\|_{\mathcal{W}^{1,\infty}((\partial\Omega)_\delta \cap U)}$$

is a bound for φ on $(\partial\Omega)_\delta \cap U$ and it is a Lipschitz constant of φ on segments $[x, y] \subset (\partial\Omega)_\delta \cap U$. First we assume that $x, y \in \partial\Omega$ such that $y \in B_\rho(x)$. Then

$$[x, y] \quad \text{belongs to} \quad C_{j_x}.$$

We can move the points of the segment $[x, y]$ parallel to the axis of the cylinder C_{j_x} to get a polygonal curve $P \subset (\partial\Omega)_\delta \cap \overline{\Omega}$ connecting finitely many points

$$x = x^0, x^1, \ldots, x^k = y \in (\partial\Omega)_\delta \cap \overline{\Omega}$$

(cf. Fig. 4.2). There is a constant $c_1 > 0$ depending merely on the largest Lipschitz constant of the γ_j such that

$$\operatorname{length}(P) \le c_1|x - y|.$$

Hence

$$|\varphi(x) - \varphi(y)| \le \sum_{j=1}^{k} |\varphi(x^{j-1}) - \varphi(x^j)| \le L \sum_{j=1}^{k} |x^{j-1} - x^j| \le c_1 L|x - y|.$$

Let now $x, y \in \partial\Omega$ be such that $y \notin B_\rho(x)$ but that x, y belong to the same component of $\overline{\Omega}$. By the convexity of the C_j there are at most m points $x_0 = x, x_1, \ldots, x_l = y$ in $\partial\Omega$ such that the closed segments

$$[x_{i-1}, x_i] \quad \text{belong to some} \quad C_j\,.$$

As above we can construct polygonal curves $P_i \subset (\partial\Omega)_\delta \cap \overline{\Omega}$ connecting x_{i-1} and x_i within $(\partial\Omega)_\delta \cap \overline{\Omega}$ to get

$$|\varphi(x_{i-1}) - \varphi(x_i)| \le c_1 L |x_{i-1} - x_i|\,.$$

Using that $y \notin B_\rho(x)$ we get

$$\begin{aligned} |\varphi(x) - \varphi(y)| &\le \sum_{i=1}^{l} |\varphi(x_{i-1}) - \varphi(x_i)| \le \sum_{i=1}^{l} c_1 L |x_{i-1} - x_i| \\ &\overset{(4.30)}{\le} \sum_{i=1}^{l} c_1 L c_0 \rho \overset{l \le m}{\le} c_0 c_1 m L |x - y|\,. \end{aligned}$$

Finally let $x, y \in \partial\Omega$ belong to different components of $\overline{\Omega}$. Then $|x - y| > \rho$ and

$$|\varphi(x) - \varphi(y)| \le |\varphi(x)| + |\varphi(y)| \le 2L \le \tfrac{2L}{\rho}|x - y|\,.$$

Summarizing all cases we use $c = 1 + \max\{1, \frac{2}{\rho}, c_1, c_0 c_1 m\}$ to get

$$\|\varphi_{|\partial\Omega}\|_{\mathrm{Lip}(\partial\Omega)} = \|\varphi_{|\partial\Omega}\|_{C(\partial\Omega)} + \mathrm{Lip}(\varphi_{|\partial\Omega}) \le cL = c\|\varphi\|_{\mathcal{W}^{1,\infty}((\partial\Omega)_\delta \cap U)}\,.$$

Since c is independent of $\delta \in (0, \rho)$ and $\varphi \in \mathcal{W}^{1,\infty}(U)$, the assumption of (1) is verified. □

4.2 Normal Measures

The linearity of the trace operator T from Proposition 3.3 hints at a linear dependence of the measures λ_F and μ_F on F. Moreover, the usual Gauss-Green formula for smooth F and regular open Ω given by

$$\operatorname{div}(\varphi F)(\Omega) = \int_{\partial\Omega} \varphi F \cdot \nu^\Omega \, d\mathcal{H}^{n-1}$$

indicates some dependence of the boundary term on the outer unit normal field ν^Ω. Therefore we are interested in more structural information about λ_F and μ_F. Here it turns out that the usage of a pointwise normal field ν^{Ω} on the boundary $\partial\Omega$ is too restrictive even if it exists (in particular if Ω contains parts of $\partial\Omega$ and if $\operatorname{div} F$ has concentrations there) and that a pointwise trace function F on the boundary might not exist. Thus we are looking for measures ν extending the usual pointwise normal fields. As a first idea we could consider some extension $\nu \in \operatorname{ba}(\Omega, \mathcal{B}(\Omega), \mathcal{L}^n)^n$ of the Radon measure $\nu^{\Omega}\mathcal{H}^{n-1}\lfloor\partial_*\Omega$ according to Proposition 2.13 for suitable sets of finite perimeter. However, by the variety of extensions, we would not get the right selection ν for a general Gauss-Green formula. If $\operatorname{div} F$ has concentrations on $\partial\Omega$, we have e.g. to take care of parts of the boundary belonging to Ω by controlling the aura of a corresponding ν. Thus we need a more careful construction of such measures than just any extension. Therefore we provide a general approach that even allows some weight on $\partial\Omega$. This way we obtain a more precise representation of the boundary term in the general Gauss-Green formula (4.1) for a large class of cases.

For an open bounded set $U \subset \mathbb{R}^n$ and for $\Omega \in \mathcal{B}(U)$, a measurable function $\chi : U \to [0, 1]$ is said to be a *good approximation* for χ_Ω if there is a sequence of functions $\chi_k \in W^{1,\infty}(U)$ such that

(1) $\chi_k : U \to [0, 1]$ is compactly supported on U for all $k \in \mathbb{N}$,
(2) $\lim_{k\to\infty} \chi_k = \chi \quad \mathcal{H}^{n-1}$-a.e. on U,
(3) $\chi_k = 1$ on $\Omega_{-\frac{1}{k}}$, $\chi_k = 0$ on $U \setminus \Omega_{\frac{1}{k}}$ for all k,
(4) $\liminf_{k\to\infty} \|D\chi_k\|_{\mathcal{L}^1(U)} < \infty$.

We call $\{\chi_k\}$ an *approximating sequence* for χ. Obviously

$$\chi = 1 \text{ on } \operatorname{int}\Omega \quad \text{and} \quad \chi = 0 \text{ on } \operatorname{ext}\Omega .$$

Notice that the χ_k cannot be taken as equivalence classes in $\mathcal{W}^{1,\infty}(U)$ and that Ω need not to be compactly contained in U for a good approximation. It might be an option for the treatment of bounded vector fields F to consider also χ_k that merely belong to $\mathcal{BV}(U)$ (recall that $\chi_k F \in \mathcal{DM}^{\infty}(U)$ in this case; cf. [11, p. 97]). Note that there is no need for such an extension in our treatment. Now we see that, aside from degenerate cases, those Ω which allow for a good approximation have to have finite perimeter in U.

Proposition 4.18 *Let $U \subset \mathbb{R}^n$ be open and bounded and let $\Omega \in \mathcal{B}(U)$. If there is a good approximation χ for χ_Ω with $\|\chi - \chi_\Omega\|_{\mathcal{L}^1(U)} = 0$, then Ω has finite perimeter in U.*

Note that $\mathcal{L}^n(\partial\Omega \cap U) = 0$ implies $\|\chi - \chi_\Omega\|_{\mathcal{L}^1(U)} = 0$.

Proof Let χ be a good approximation for χ_Ω with corresponding approximating sequence $\{\chi_k\} \subset W^{1,\infty}(U)$. Since Ω is bounded and every $\mathcal{H}^{n-1}$-null set is $\mathcal{L}^n$-null set,

$$\chi_k \xrightarrow{L^1} \chi_\Omega .$$

By the lower semicontinuity of the total variation of BV functions,

$$|D\chi_\Omega|(U) \le \liminf_{k\to\infty} \|D\chi_k\|_{\mathcal{L}^1} < \infty .$$

Hence $\chi_\Omega \in BV(U)$ and the assertion follows. □

Let us now demonstrate that good approximations provide measures.

Theorem 4.19 *Let $U \subset \mathbb{R}^n$ be open and bounded, let $\Omega \in \mathcal{B}(U)$, and let χ be a good approximation for χ_Ω with approximating sequence $\{\chi_k\}$. Then there is an associated measure $\nu \in \mathrm{ba}\big(U, \mathcal{B}(U), \mathcal{L}^n\big)^n$ such that*

$$\operatorname{core}\nu \subset \partial\Omega\,, \quad |\nu|(U) = \|\nu\| \le \liminf_{k\to\infty} \|D\chi_k\|_{\mathcal{L}^1(U)}\,,$$

$$|\nu|(B) \le \limsup_{k\to\infty} \|D\chi_k\|_{\mathcal{L}^1(B)} \quad \text{for all} \quad B \in \mathcal{B}(U)\,, \tag{4.31}$$

$$A = \bigcup_{k\ge k_0} \operatorname{supp}(D\chi_k) \tag{4.32}$$

is an aura of ν for each $k_0 \in \mathbb{N}$, and for any $\varphi \in \mathcal{L}^\infty(U, \mathbb{R}^n)$ there is a subsequence $\{\chi_{k'}\}$ with

$$\lim_{k'\to\infty} \int_U \varphi \cdot D\chi_{k'}\, d\mathcal{L}^n = -\fint_{\partial\Omega} \varphi\, d\nu\,. \tag{4.33}$$

Moreover, ν is a trace on $\partial\Omega$ over $\mathcal{L}^\infty(U, \mathbb{R}^n)$. If $\|\chi - \chi_\Omega\|_{\mathcal{L}^1(U)} = 0$, then

$$\fint_{\partial\Omega} \varphi\, d\nu = \int_{\partial_*\Omega\cap U} \varphi \cdot \nu^\Omega\, d\mathcal{H}^{n-1} \quad \text{for all} \quad \varphi \in C_c(U, \mathbb{R}^n) \tag{4.34}$$

and

$$|\nu|(B) \ge \big(\mathcal{H}^{n-1}\llcorner(\partial^*\Omega \cap U)\big)(B) \quad \text{for all open} \quad B \subset U\,. \tag{4.35}$$

If $U = \Omega$ with Lipschitz boundary, then

$$\fint_{\partial\Omega} \varphi\, d\nu = \int_{\partial\Omega} \varphi \cdot \nu^\Omega\, d\mathcal{H}^{n-1} \quad \text{for all} \quad \varphi \in C(\overline{\Omega}, \mathbb{R}^n) \tag{4.36}$$

and we have (4.35) *with $\partial\Omega$ instead of $\partial^*\Omega \cap U$ for all relatively open $B \subset \overline{\Omega}$.*

We call ν (outward) *normal measure* of Ω. The proof shows that ν is a weak* cluster point of the measures $-D\chi_k\mathcal{L}^n$ and thus, it might be not unique in general. However it is uniquely determined by the subsequences entering (4.33). Below we provide examples showing that, for given Ω, there are different normal measures based on different auras. In particular we consider normal measures with aura being completely inside or completely outside Ω where, in both cases, the vectors $\nu(B)$ are directed outward for small balls B intersecting the boundary $\partial\Omega$. But notice that we always have the same right hand side in (4.34) for continuous φ. If Ω has Lipschitz boundary, then $\partial_*\Omega = \partial\Omega$ (cf. [32, p. 50]). The inequality in (4.35) can be strict as e.g. in the simple case of an open ball $\Omega = B \Subset U$ and $\nu = \nu_{\text{int}}$ from Example 4.21 below where

$$|\nu|(B) = \mathcal{H}^{n-1}(\partial\Omega) > (\mathcal{H}^{n-1}\llcorner\partial\Omega)(B) = 0\,.$$

Proof Since χ is a good approximation, the measures $\nu_k := -D\chi_k\mathcal{L}^n$ with norm $\|\nu_k\| = \|D\chi_k\|_{\mathcal{L}^1(U)}$ have a subsequence $\nu_{k'}$ that is bounded in $\mathcal{L}^\infty(U,\mathbb{R}^n)^*$ and satisfies

$$\lim_{k'\to\infty} \|D\chi_{k'}\|_{\mathcal{L}^1(U)} = \liminf_{k\to\infty} \|D\chi_k\|_{\mathcal{L}^1(U)}\,.$$

This subsequence has a weak* cluster point in $\mathcal{L}^\infty(U,\mathbb{R}^n)^*$ that we can identify with some

$$\nu \in \text{ba}\big(U, \mathcal{B}(U), \mathcal{L}^n\big)^n \quad \text{where} \quad |\nu|(U) = \|\nu\| \le \liminf_{k\to\infty} \|D\chi_k\|_{\mathcal{L}^1(U)}\,.$$

Consequently we have (4.33) where the subsequence $\{\chi_{k'}\}$ might depend on φ. Hence $\int_U \varphi\, d\nu = 0$ for all φ vanishing on $(\partial\Omega)_\delta$ with some $\delta > 0$. Thus $\text{core}\,\nu \subset \partial\Omega$ and ν is a trace on $\partial\Omega$ over $\mathcal{L}^\infty(U,\mathbb{R}^n)$. Taking φ vanishing on $U \setminus A$, we readily get from (4.33) that A is an aura of ν. For $B \in \mathcal{B}(U)$ and $\varepsilon > 0$ there is some $\varphi \in \mathcal{L}^\infty(U,\mathbb{R}^n)$ with $\|\varphi\|_\infty \le 1$ and a subsequence $\chi_{k'}$ such that

$$\begin{aligned} |\nu|(B) - \varepsilon &\le \fint_{\partial\Omega} \chi_B\varphi\, d\nu = \lim_{k'\to\infty} \int_U \chi_B\varphi \cdot D\chi_{k'}\, d\mathcal{L}^n \\ &\le \limsup_{k'\to\infty} \int_B |D\chi_{k'}|\, d\mathcal{L}^n \le \limsup_{k\to\infty} \|D\chi_k\|_{\mathcal{L}^1(B)} \end{aligned}$$

(cf. (2.9)).

Let now $\|\chi - \chi_\Omega\|_{\mathcal{L}^1(U)} = 0$. Then $\chi_k \to \chi_\Omega$ in $L^1(U)$ and Ω has finite perimeter in U by Proposition 4.18. For $\varphi \in C_c^\infty(U,\mathbb{R}^n)$ the definition of weak derivatives gives

$$\int_U \varphi \cdot D\chi_k\, d\mathcal{L}^n = -\int_U \chi_k \,\text{div}\,\varphi\, d\mathcal{L}^n\,. \tag{4.37}$$

By dominated convergence and the divergence theorem

$$\lim_{k\to\infty} \int_U \chi_k\,\text{div}\,\varphi\, d\mathcal{L}^n = \int_\Omega \text{div}\,\varphi\, d\mathcal{L}^n = \int_{\partial_*\Omega\cap U} \varphi\cdot\nu^\Omega\, d\mathcal{H}^{n-1} \tag{4.38}$$

(cf. [32, p. 141] and notice that φ is Lipschitz continuous due to its compact support). Using (4.33) for the left hand side in (4.37), we obtain

$$\fint_{\partial\Omega} \varphi \, d\nu = \int_{\partial_*\Omega\cap U} \varphi \cdot \nu^{\Omega} \, d\mathcal{H}^{n-1}$$

for all $\varphi \in C_c^{\infty}(U, \mathbb{R}^n)$. By uniform approximation we get this identity even for all $\varphi \in C_c(U, \mathbb{R}^n)$.

Now let $B \subset U$ be open and recall that

$$D\chi_{\Omega} = \nu^{\Omega}\mathcal{H}^{n-1} \llcorner (\partial^*\Omega \cap U), \quad |D\chi_{\Omega}| = \mathcal{H}^{n-1} \llcorner (\partial^*\Omega \cap U)$$

(cf. [22, pp. 169, 205]). Then, with (2.9),

$$\begin{aligned}
|\nu|(B) &= \sup_{\substack{\varphi\in\mathcal{L}^{\infty}(U,\mathbb{R}^n)\\ \|\varphi\|_{\infty}\le 1}} \int_B \varphi \, d\nu \\
&\ge \sup_{\substack{\varphi\in C_c^1(B,\mathbb{R}^n)\\ \|\varphi\|_{\infty}\le 1}} \int_B \varphi \, d\nu = \sup_{\substack{\varphi\in C_c^1(B,\mathbb{R}^n)\\ \|\varphi\|_{\infty}\le 1}} \int_U \varphi \, d\nu \\
&\overset{(4.34)}{=} \sup_{\substack{\varphi\in C_c^1(B,\mathbb{R}^n)\\ \|\varphi\|_{\infty}\le 1}} \int_{\partial_*\Omega\cap U} \varphi \cdot \nu^{\Omega} \, d\mathcal{H}^{n-1} \\
&= \sup_{\substack{\varphi\in C_c^1(B,\mathbb{R}^n)\\ \|\varphi\|_{\infty}\le 1}} \int_U \varphi \, dD\chi_{\Omega} \\
&= \sup_{\substack{\varphi\in C_c^1(B,\mathbb{R}^n)\\ \|\varphi\|_{\infty}\le 1}} \int_B \chi_{\Omega} \operatorname{div} \varphi \, d\mathcal{L}^n \\
&= |D\chi_{\Omega}|(B) = \big(\mathcal{H}^{n-1} \llcorner (\partial^*\Omega \cap U)\big)(B) .
\end{aligned}$$

(cf. also [22, p. 170] for the second last equality).

If $U = \Omega$ with Lipschitz boundary, then $\|\chi - \chi_{\Omega}\|_{\mathcal{L}^1(U)} = 0$ and we argue similar as above. For (4.36) we consider

$$\varphi \in C^{\infty}(\Omega, \mathbb{R}^n) \cap C^1(\overline{\Omega}, \mathbb{R}^n) .$$

Then we get (4.37), since χ_k has compact support in Ω, and in (4.38) we use the divergence theorem from [32, p. 168] to get

$$\int_{\Omega} \operatorname{div} \varphi \, d\mathcal{L}^n = \int_{\partial\Omega} \varphi \cdot \nu^{\Omega} \, d\mathcal{H}^{n-1}$$

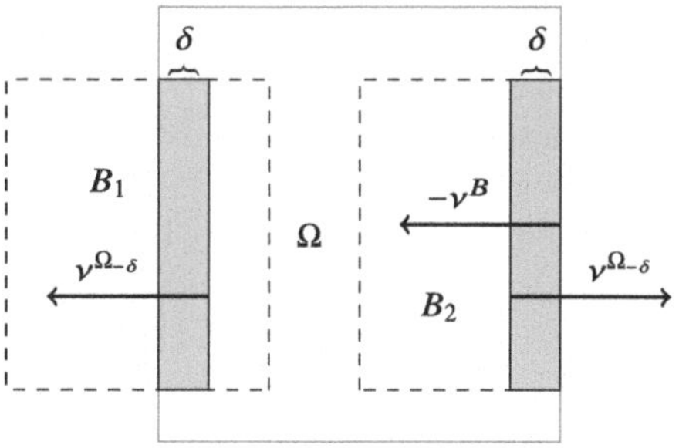

Fig. 4.3 Contributions to $\nu(B)$ for two versions of B

(notice that φ is locally Lipschitz continuous on $\overline{\Omega}$). For the adaption of (4.35) we extend ν by zero on $\mathbb{R}^n$, we argue as above for arbitrary open $B \subset \mathbb{R}^n$ with some enlarged U, and we use that $\partial^*\Omega = \partial\Omega$ (cf. [32, p. 50]). □

Proposition 4.20 *Let $U \subset \mathbb{R}^n$ be open and bounded, let $\Omega \in \mathcal{B}(U)$ be a set of finite perimeter, and let χ be a good approximation for χ_Ω with approximating sequence χ_k and associated normal measure $\nu \in \mathrm{ba}\big(U, \mathcal{B}(U), \mathcal{L}^n\big)^n$. Then for every $B \in \mathcal{B}(U)$ with finite perimeter there exists an $\mathcal{L}^1$-null set $N \subset \mathbb{R}$ and some $\tilde{\delta} > 0$ such that for all $\delta \in (0, \tilde{\delta}) \setminus N$*

$$\begin{aligned}
\nu(B) &= -\lim_{k\to\infty} \int_B D\chi_k \, d\mathcal{L}^n \\
&= -\int_{U\cap\overline{\Omega}\cap\partial_*(B\cap\Omega^\delta)} \chi\, \nu^{B\cap\Omega^\delta} d\mathcal{H}^{n-1} \quad \text{with} \quad \Omega^\delta = \Omega_\delta \setminus \Omega_{-\delta} \\
&= \int_{\partial_* B\cap\partial\Omega\cap U} -\chi\, \nu^B \, d\mathcal{H}^{n-1} \\
&\quad + \lim_{\substack{\delta\downarrow 0 \\ \delta\notin N}} \int_{(\mathrm{int}_* B)\cap\partial_*\Omega_{-\delta}\cap U} \nu^{\Omega_{-\delta}} d\mathcal{H}^{n-1} .
\end{aligned} \tag{4.39}$$

The representation of $\nu(B)$ is illustrated in Fig. 4.3 for two simple cases. For B_1 the first integral in (4.39) vanishes and thus, $\nu(B)$ is directed as ν^Ω on $\partial\Omega \cap B_1$ with length $\mathcal{H}^{n-1}(\partial\Omega \cap B_1)$. For B_2 we distinguish two cases. If $\chi = 0$ on $\partial\Omega$, then we have a similar situation as for B_1 and $-\nu^B$ in the figure does not apply. If $\chi = 1$ on $\partial\Omega$, then $\nu(B) = 0$, since the two contributions in (4.39) cancel each other. Below we discuss this situation in more detail for several examples.

Proof Since $\nu <<^w \mathcal{L}^n$ and since B differs from $\mathrm{int}_* B$ only by an $\mathcal{L}^n$-null set (cf. [22, p. 222]), we have $\nu(B) = \nu(\mathrm{int}_* B)$. Thus we can essentially work with $\tilde{B} := \mathrm{int}_* B$, but in integrals with $\mathcal{L}^n$-measure we can replace it with the original B. We proceed analogously with $\tilde{\Omega}^\delta := \mathrm{int}_* \Omega^\delta$. Clearly $\partial_* B = \partial_* \tilde{B}$. By $\mathrm{int}_*(B \cap \Omega^\delta) = \tilde{B} \cap \tilde{\Omega}^\delta$ (cf. [32, p. 50]), we also have $\partial_*(B \cap \Omega^\delta) = \partial_*(\tilde{B} \cap \tilde{\Omega}^\delta)$. The coarea formula implies that Ω^δ and thus also $\tilde{\Omega}^\delta$ has finite perimeter for a.e. $\delta > 0$. Therefore $\tilde{B} \cap \tilde{\Omega}^\delta$ has finite perimeter too (cf. [31, p. 130]). Notice that $\tilde{\Omega}^\delta$ might not be a subset of

U and, though we have assumed $B \subset U$, also $\tilde{B}$ might not be a subset of U. But $\tilde{B} \cap \tilde{\Omega}^\delta$ has finite perimeter in U for a.e. $\delta > 0$.

We now consider the approximating sequence $\chi_k \in W^{1,\infty}(U)$ for χ and we choose some $\varphi \in C^1(U, \mathbb{R}^n)$. Then $\chi_k \varphi$ is Lipschitz continuous with compact support on U, since it is locally Lipschitz continuous. Hence, using

$$U \cap \partial_*(\tilde{B} \cap \tilde{\Omega}^\delta \cap U) = U \cap \partial_*(\tilde{B} \cap \tilde{\Omega}^\delta)$$

and a general version of the Gauss-Green formula, we obtain for a.e. $\delta > 0$

$$\int_{B\cap\Omega^\delta} \varphi \cdot D\chi_k \, d\mathcal{L}^n = -\int_{B\cap\Omega^\delta} \chi_k \operatorname{div} \varphi \, d\mathcal{L}^n + \int_{U\cap\partial_*(\tilde{B}\cap\tilde{\Omega}^\delta)} \chi_k \varphi \cdot \nu^{\tilde{B}\cap\tilde{\Omega}^\delta} \, d\mathcal{H}^{n-1}$$

(cf. [32, pp. 51, 141] and [25, 4.5.6]). Since $\chi_k \to \chi$ $\mathcal{H}^{n-1}$-a.e. on U, by dominated convergence

$$\lim_{k\to\infty} \int_{B\cap\Omega^\delta} \varphi \cdot D\chi_k \, d\mathcal{L}^n = -\int_{B\cap\Omega^\delta} \chi \operatorname{div} \varphi \, d\mathcal{L}^n + \int_{U\cap\partial_*(\tilde{B}\cap\tilde{\Omega}^\delta)} \chi \varphi \cdot \nu^{\tilde{B}\cap\tilde{\Omega}^\delta} \, d\mathcal{H}^{n-1}$$

for all $\varphi \in C(U, \mathbb{R}^n)$. Let us choose φ to equal a constant vector $a \in \mathbb{R}^n$ on U. By core $\nu \subset \partial\Omega$ and by (4.33), where we do not need a subsequence due to the previous equation,

$$\begin{aligned}
\lim_{k\to\infty} \int_{B\cap\Omega^\delta} a \cdot D\chi_k \, d\mathcal{L}^n &= \lim_{k\to\infty} \int_U \chi_{B\cap\Omega^\delta} \, a \cdot D\chi_k \, d\mathcal{L}^n \\
&= -a \cdot \fint_{\partial\Omega} \chi_{B\cap\Omega^\delta} \, d\nu = -a \cdot \nu(B \cap \Omega^\delta) \\
&= -a \cdot \nu(B) \quad \text{for all} \quad a \in \mathbb{R}^n .
\end{aligned}$$

The arbitrariness of $a \in \mathbb{R}^n$ implies

$$\nu(B) = -\lim_{k\to\infty} \int_{B\cap\Omega^\delta} D\chi_k \, d\mathcal{L}^n = -\int_{U\cap\partial_*(\tilde{B}\cap\tilde{\Omega}^\delta)} \chi \, \nu^{\tilde{B}\cap\tilde{\Omega}^\delta} \, d\mathcal{H}^{n-1} \tag{4.40}$$

for a.e. $\delta > 0$. By $\operatorname{supp} D\chi_k \subset (\partial\Omega)_{1/k}$ we can omit Ω^δ in the first integral and by $\operatorname{supp} \chi \subset \overline{\Omega}$ we can restrict the second integral to $\overline{\Omega}$. Moreover the previous arguments show that (4.40) is also valid with B, Ω^δ instead of $\tilde{B}$, $\tilde{\Omega}^\delta$.

Since $\tilde{B}$ and $\tilde{\Omega}^\delta$ agree with their measure theoretic interior,

$$(\tilde{B} \cap \partial_* \tilde{\Omega}^\delta) \cup (\partial_* \tilde{B} \cap \tilde{\Omega}^\delta) \subset \partial_*(\tilde{B} \cap \tilde{\Omega}^\delta)$$
$$\subset (\tilde{B} \cap \partial_* \tilde{\Omega}^\delta) \cup (\partial_* \tilde{B} \cap \tilde{\Omega}^\delta) \cup (\partial_* \tilde{B} \cap \partial_* \tilde{\Omega}^\delta)$$

(cf. [32, p. 52]). By $\mathcal{L}^n(\partial_* \tilde{B}) = 0$, the coarea formula implies

$$\mathcal{H}^{n-1}(\partial_* \tilde{B} \cap \partial_* \tilde{\Omega}^\delta) = 0 \quad \text{for a.e. } \delta > 0 .$$

Therefore $\partial_*(\tilde{B} \cap \tilde{\Omega}^\delta)$ differs from $(\tilde{B} \cap \partial_* \tilde{\Omega}^\delta) \cup (\partial_* \tilde{B} \cap \tilde{\Omega}^\delta)$ only by an $\mathcal{H}^{n-1}$-null set for these $\delta > 0$. Furthermore

$$\nu^{\tilde{B} \cap \tilde{\Omega}^\delta} = \nu^{\tilde{B}} \quad \text{on} \quad \partial^*(\tilde{B} \cap \tilde{\Omega}^\delta) \cap (\partial^* \tilde{B} \cap \tilde{\Omega}^\delta) \quad \text{for all } \delta > 0 ,$$

since for such x (with dens_x from Example 2.4)

$$\operatorname{dens}_x(\tilde{B} \cap \tilde{\Omega}^\delta) = \operatorname{dens}_x \tilde{B} = \tfrac{1}{2}$$

and, by $\tilde{B} \cap \tilde{\Omega}^\delta \subset \tilde{B}$, both sets have to generate the same half-space determining their normal (cf. [2, p. 157/158]). Analogously we get

$$\nu^{\tilde{B} \cap \tilde{\Omega}^\delta} = \nu^{\tilde{\Omega}^\delta} \quad \text{on} \quad \partial^*(\tilde{B} \cap \tilde{\Omega}^\delta) \cap (\tilde{B} \cap \partial^* \tilde{\Omega}^\delta) \quad \text{for all } \delta > 0 .$$

Using that the reduced boundary agrees $\mathcal{H}^{n-1}$-a.e. with the measure theoretic one, that $\partial_* B = \partial_* \tilde{B}$, and that $\partial_* \Omega^\delta = \partial_* \tilde{\Omega}^\delta$, we obtain from (4.40) that

$$\nu(B) = -\int_{\partial_* B \cap \tilde{\Omega}^\delta \cap U} \chi \nu^B \, d\mathcal{H}^{n-1} - \int_{\tilde{B} \cap \partial_* \Omega^\delta \cap U} \chi \nu^{\Omega^\delta} \, d\mathcal{H}^{n-1} \tag{4.41}$$

for all $\delta > 0$ despite an $\mathcal{L}^1$-null set N. Since

$$A \to \int_{\partial_* B \cap A \cap U} \chi \nu^B \, d\mathcal{H}^{n-1}$$

is a σ-measure on $\partial_* B \cap U$ and $\bigcap_{\delta > 0} \tilde{\Omega}^\delta = \partial \Omega$,

$$\lim_{\substack{\delta \downarrow 0 \\ \delta \notin N}} \int_{\partial_* B \cap \tilde{\Omega}^\delta \cap U} \chi \nu^B \, d\mathcal{H}^{n-1} = \int_{\partial_* B \cap \partial \Omega \cap U} \chi \nu^B \, d\mathcal{H}^{n-1} .$$

For the second integral in (4.41) we use $\partial_*\Omega^\delta = \partial_*(\Omega_\delta \setminus \Omega_{-\delta})$,

$$\operatorname{supp}\chi \cap \partial_*\Omega^\delta = \partial_*\Omega_{-\delta}\,, \quad \tilde{B} = \operatorname{int}_* B, \quad \nu^{\Omega^\delta} = -\nu^{\Omega_{-\delta}} \text{ on } \partial_*\Omega_{-\delta}$$

to get the assertion. □

Let us now provide some important normal measures ν of Ω. Some of the measures use the distance function $\operatorname{dist}_{\partial\Omega}$ for the approximating sequence χ_k and require some boundedness of $\mathcal{H}^{n-1}(\partial\Omega_\delta)$ for δ near zero. Notice that for any open bounded Ω

$$\operatorname{Per}(\Omega_\delta) \le \mathcal{H}^{n-1}(\partial\Omega_\delta) < \infty \quad \text{for all} \quad \delta \ne 0$$

(cf. [29, Theorem 3]), but there are open bounded sets Ω of finite perimeter where $\mathcal{H}^{n-1}(\partial\Omega_\delta)$ scales as δ^{-s} for $s \in (0,1)$ (cf. [29, Theorem 1]). The other examples are applicable to sets of finite perimeter $\Omega \Subset U$ and use mollifications of χ_Ω for the approximating sequence χ_k.

As preparation let $\Omega \Subset U$ have finite perimeter and consider

$$\psi_k := \chi_\Omega * \eta_{\frac{1}{k}} \tag{4.42}$$

with the standard mollifier η_ε supported on $B_\varepsilon(0)$. Then, for large k,

$$\psi_k \in C_c^\infty(U,[0,1])\,, \quad \psi_k = 1 \text{ on } \Omega_{-\frac{1}{k}}\,, \quad \psi_k = 0 \text{ on } U \setminus \Omega_{\frac{1}{k}}\,,$$

$$\psi_k \to \tfrac{1}{2}\chi_{\partial_*\Omega} + \chi_{\operatorname{int}_*\Omega} \quad \mathcal{H}^{n-1}\text{-a.e. on } U \tag{4.43}$$

(cf. [2, pp. 164, 173, 175]). From [2, pp. 41–42, 118] we obtain for large k

$$D\psi_k(x) = \big(D\chi_\Omega * \eta_{\frac{1}{k}}\big)(x) := \int_U \eta_{\frac{1}{k}}(x-y)\, dD\chi_\Omega(y) \quad \text{for } x \in \Omega\,.$$

Then one has the weak* limits

$$D\psi_k \mathcal{L}^n \overset{*}{\rightharpoonup} D\chi_\Omega \quad \text{and} \quad |D\psi_k|\mathcal{L}^n \overset{*}{\rightharpoonup} |D\chi_\Omega| \tag{4.44}$$

in the sense of Radon measures. For any Borel set $B \subset U$ and k large, we can use $\operatorname{supp} D\psi_k \subset \Omega_{\frac{1}{k}} \Subset U$ and $\partial_*\Omega \subset \partial\Omega$ to get

$$\begin{aligned} \|D\psi_k\|_{\mathcal{L}^1(B)} &= \int_{\Omega_{1/k}\cap B} |D\psi_k|\, d\mathcal{L}^n \\ &\le |D\chi_\Omega|(\Omega_{\frac{2}{k}} \cap B_{\frac{1}{k}}) \;=\; \mathcal{H}^{n-1}(\partial_*\Omega \cap B_{\frac{1}{k}}) \end{aligned} \tag{4.45}$$

(cf. [2, p. 42], [32, pp. 50, 138] or also [31, pp. 49, 129]). If $|D\chi_\Omega|(\partial B) = 0$, then

$$\lim_{k\to\infty} |D\psi_k|(B) = |D\chi_\Omega|(B) \tag{4.46}$$

(cf. [22, p. 54]). Obviously $\psi_k \in W^{1,\infty}(U)$ and also the truncations

$$\tilde\chi_k^{\text{int}} := (2\psi_k - 1)^+ , \quad \tilde\chi_k^{\text{ext}} := -(2\psi_k - 1)^- + 1$$

(where $(\ldots)^\pm$ is the positive/negative part) belong to $W^{1,\infty}(U)$ with

$$D\tilde\chi_k^{\text{int}} = \begin{cases} 2D\psi_k \ \mathcal{L}^n\text{-a.e. on } \{\psi_k > \frac{1}{2}\}, \\ \quad 0 \quad \mathcal{L}^n\text{-a.e. on } \{\psi_k \le \frac{1}{2}\} \end{cases}$$

and $D\tilde\chi_k^{\text{ext}}$ analogously (cf. [22, p. 130]). Since the sets Ω and $\text{int}_*\Omega$ differ merely by an $\mathcal{L}^n$-negligible set (cf. [22, p. 222]), we can use (3.1) with $\varphi \in C_c^1(U)$ to get

$$\begin{aligned} \lim_{k\to\infty} \int_U \varphi D\tilde\chi_k^{\text{int}} \, d\mathcal{L}^n &= - \lim_{k\to\infty} \int_U \tilde\chi_k^{\text{int}} \operatorname{div} \varphi \, d\mathcal{L}^n = - \int_U \chi_{\text{int}_*\Omega} \operatorname{div} \varphi \, d\mathcal{L}^n \\ &= - \int_U \chi_\Omega \operatorname{div} \varphi \, d\mathcal{L}^n = \int_U \varphi \, dD\chi_\Omega \end{aligned}$$

and analogously for $\tilde\chi_k^{\text{ext}}$. Since $C_c^1(U)$ is dense in $C_c(U)$, we have

$$D\tilde\chi_k^{\text{int}} \mathcal{L}^n \overset{*}{\rightharpoonup} D\chi_\Omega \quad \text{and} \quad D\tilde\chi_k^{\text{ext}} \mathcal{L}^n \overset{*}{\rightharpoonup} D\chi_\Omega .$$

From (4.45) we get for any Borel set $B \subset U$

$$\begin{aligned} \|D\tilde\chi_k^{\text{int}}\|_{\mathcal{L}^1(B)} &= \int_{B\cap\{\psi_k > \frac{1}{2}\}} 2|D\psi_k| \, d\mathcal{L}^n \le 2 \int_B |D\psi_k| \, d\mathcal{L}^n \\ &\le 2\mathcal{H}^{n-1}(\partial_*\Omega \cap B_{\frac{1}{k}}) \end{aligned} \tag{4.47}$$

and analogously

$$\|D\tilde\chi_k^{\text{ext}}\|_{\mathcal{L}^1(B)} \le 2\mathcal{H}^{n-1}(\partial_*\Omega \cap B_{\frac{1}{k}}) . \tag{4.48}$$

Let us now introduce some special examples of normal measures as provided by Theorem 4.19. These can be basically applied to the same set Ω. However they differ in the way how the boundary is taken into account or in the underlying construction.

Example 4.21 (Interior Normal Measures) Let $U \subset \mathbb{R}^n$ be an open and bounded set and let $\Omega \in \mathcal{B}(U)$. We call a normal measure *interior normal measure* of Ω if the related good approximation χ has the form

$$\chi = \chi_{\operatorname{int}\Omega} \quad \text{or} \quad \chi = \chi_{\operatorname{int}_*\Omega} .$$

For a first case we assume that $\Omega \Subset U$, that there is a sequence $\delta_k \downarrow 0$ with

$$\lim_{k\to\infty} \fint_{(0,\delta_k)} \mathcal{H}^{n-1}(\partial\Omega_{-\delta})\, d\delta < \infty , \tag{4.49}$$

and we consider

$$\chi_k^{\text{int}} := \chi_{\overline{\Omega_{-\delta_k}}} + \chi_{(\operatorname{int}\Omega\setminus\overline{\Omega_{-\delta_k}})} \frac{\operatorname{dist}_{\partial\Omega}}{\delta_k} \in W^{1,\infty}(U) . \tag{4.50}$$

We can suppose that $\delta_k \le \frac{1}{k}$ and we obviously have that

$$\chi_k^{\text{int}} \to \chi_{\operatorname{int}\Omega} \quad \text{pointwise on } U .$$

By the coarea formula (cf. [22, p.117]),

$$\begin{aligned} \|D\chi_k^{\text{int}}\|_{\mathcal{L}^1(U)} &= \int_{(0,\delta_k)} \int_{\partial\Omega_{-\delta}} |D\chi_k^{\text{int}}|\, d\mathcal{H}^{n-1}\, d\delta \\ &= \int_{(0,\delta_k)} \frac{\mathcal{H}^{n-1}(\partial\Omega_{-\delta})}{\delta_k} d\delta \\ &= \fint_{(0,\delta_k)} \mathcal{H}^{n-1}(\partial\Omega_{-\delta})\, d\delta \end{aligned}$$

(notice that $|D \operatorname{dist}_{\partial\Omega}(x)| = 1$ $\mathcal{L}^n$-a.e. outside $\partial\Omega$, cf. (4.3)) Hence we obtain a good approximation for χ_Ω and a related interior normal measure by

$$\chi^{\text{int}} := \chi_{\operatorname{int}\Omega} \quad \text{and} \quad \nu^{\text{int}} \in \operatorname{ba}\big(U, \mathcal{B}(U), \mathcal{L}^n\big)^n .$$

Notice that any $A = (\partial\Omega)_{\delta_k} \cap \operatorname{int}\Omega$ is an aura of ν^{int} by (4.32).

As alternative approximating sequence for $\chi_{\operatorname{int}\Omega}$ we set $\delta_k' := \frac{\delta_k}{2}$ and take

$$\chi_k^{\text{intc}} := \chi_{\overline{\Omega_{-2\delta_k'}}} + \chi_{(\Omega_{-\delta_k'}\setminus\overline{\Omega_{-2\delta_k'}})} \frac{\operatorname{dist}_{\partial(\Omega_{-\delta_k'})}}{\delta_k'} \in W^{1,\infty}(U)$$

with $0 < \delta_k \le \frac{1}{k}$ as far as

$$\lim_{k\to\infty} \fint_{(\delta_k',2\delta_k')} \mathcal{H}^{n-1}(\partial\Omega_{-\delta})\, d\delta < \infty$$

("intc" indicates compact support of $D\chi_k^{\text{intc}}$ on $\operatorname{int}\Omega$). In this case we can drop the requirement $\Omega \Subset U$ and obtain a second interior normal measure related to the good approximation $\chi_{\operatorname{int}\Omega}$ of χ_Ω by

$$\chi^{\text{intc}} := \chi_{\operatorname{int}\Omega} \quad \text{and} \quad \nu^{\text{intc}} \in \operatorname{ba}\big(U, \mathcal{B}(U), \mathcal{L}^n\big)^n .$$

Though the measures ν^{intc} and ν^{int} slightly differ, this is not relevant for the values of the corresponding integrals in the considered Gauss-Green formulas.

For a second case we assume that $\Omega \Subset U$ has finite perimeter and we consider

$$\tilde{\chi}_k^{\text{int}} := (2\psi_k - 1)^+$$

with the mollification ψ_k from (4.42). Then, by (4.43),

$$\tilde{\chi}_k^{\text{int}} \to \chi_{\operatorname{int}_*\Omega} \quad \mathcal{H}^{n-1}\text{-a.e. on } U.$$

Obviously $\tilde{\chi}_k^{\text{int}} \in W^{1,\infty}(U)$ and, by (4.47),

$$\|D\tilde{\chi}_k^{\text{int}}\|_{\mathcal{L}^1(U)} = \int_{\{\psi_k > \frac{1}{2}\}} 2|D\psi_k|\, d\mathcal{L}^n \le 2\mathcal{H}^{n-1}(\partial_*\Omega). \tag{4.51}$$

This way we get a good approximation for χ_Ω and a related interior normal measure by

$$\tilde{\chi}^{\text{int}} := \chi_{\operatorname{int}_*\Omega} \quad \text{and} \quad \tilde{\nu}^{\text{int}} \in \operatorname{ba}\big(U, \mathcal{B}(U), \mathcal{L}^n\big)^n .$$

Example 4.22 (Exterior Normal Measures) Let $U \subset \mathbb{R}^n$ be an open and bounded set and let $\Omega \in \mathcal{B}(U)$. We call a normal measure *exterior normal measure* of Ω if the related good approximation χ has the form

$$\chi = \chi_{\overline{\Omega}} \quad \text{or} \quad \chi = \chi_{\partial_*\Omega \cup \operatorname{int}_*\Omega} .$$

First we assume that $\Omega \Subset U$, that there is a sequence $\delta_k \downarrow 0$ with

$$\lim_{k\to\infty} \fint_{(0,\delta_k)} \mathcal{H}^{n-1}(\partial\Omega_\delta)\, d\delta < \infty , \tag{4.52}$$

and we consider

$$\chi_k^{\text{ext}} = \chi_{\overline{\Omega}} + \chi_{(\Omega_{\delta_k} \setminus \overline{\Omega})} \frac{\operatorname{dist}_{\partial(\Omega_{\delta_k})}}{\delta_k} \in W^{1,\infty}(U) .$$

We again suppose that $\delta_k \le \frac{1}{k}$ and we have that

$$\chi_k^{\text{ext}} \to \chi_{\overline{\Omega}} \quad \text{pointwise on } U\,.$$

As above,

$$\|D\chi_k^{\text{ext}}\|_{\mathcal{L}^1(U)} = \fint_{(0,\delta_k)} \mathcal{H}^{n-1}(\partial\Omega_\delta)\,d\delta\,.$$

Therefore we obtain a good approximation for χ_Ω and a related exterior normal measure by

$$\chi^{\text{ext}} := \chi_{\overline{\Omega}} \quad \text{and} \quad \nu^{\text{ext}} \in \text{ba}\big(U, \mathcal{B}(U), \mathcal{L}^n\big)^n\,.$$

Now any $A = (\partial\Omega)_{\delta_k} \cap \text{ext}\,\Omega$ is an aura of ν^{ext} by (4.32). Similar as in the previous example we can construct some other exterior normal measure ν^{extc} related to an approximating sequence $\{\chi_k^{\text{extc}}\}$ for $\chi^{\text{ext}} = \chi_{\overline{\Omega}}$ where $\chi_k^{\text{extc}} = 1$ on Ω_{δ_k}.

For a second example we assume that $\Omega \Subset U$ has finite perimeter and we consider

$$\tilde{\chi}_k^{\text{ext}} := -(2\psi_k - 1)^- + 1$$

with ψ_k from (4.42). Then, by (4.43),

$$\tilde{\chi}_k^{\text{ext}} \to \chi_{\partial_*\Omega \cup \text{int}_*\Omega} \quad \mathcal{H}^{n-1}\text{-a.e. on } U.$$

Clearly $\tilde{\chi}_k^{\text{ext}} \in W^{1,\infty}(U)$ and, by (4.48),

$$\|D\tilde{\chi}_k^{\text{ext}}\|_{\mathcal{L}^1(U)} \le 2\mathcal{H}^{n-1}(\partial_*\Omega)\,.$$

Hence we get a good approximation for χ_Ω and a related exterior normal measure by

$$\tilde{\chi}^{\text{ext}} := \chi_{\partial_*\Omega \cup \text{int}_*\Omega} \quad \text{and} \quad \tilde{\nu}^{\text{ext}} \in \text{ba}\big(U, \mathcal{B}(U), \mathcal{L}^n\big)^n\,.$$

Example 4.23 (Symmetric Normal Measures) Let $U \subset \mathbb{R}^n$ be an open and bounded set and let $\Omega \in \mathcal{B}(U)$. We call a normal measure *symmetric normal measure* of Ω if the related good approximation χ has the form

$$\chi = \tfrac{1}{2}\chi_{\partial\Omega} + \chi_{\text{int}\,\Omega} \quad \text{or} \quad \chi = \tfrac{1}{2}\chi_{\partial_*\Omega} + \chi_{\text{int}_*\Omega}\,.$$

Let us first assume that $\Omega \Subset U$, that there is a sequence $\delta_k \downarrow 0$ with

$$\lim_{k\to\infty} \fint_{(-\delta_k,\delta_k)} \mathcal{H}^{n-1}(\partial\Omega_\delta)\, d\delta < \infty\,, \tag{4.53}$$

and let us take

$$\chi_k^{\mathrm{sym}} = \chi_{\overline{\Omega_{-\delta_k}}} + \chi_{(\Omega_{\delta_k}\setminus\overline{\Omega_{-\delta_k}})} \frac{\mathrm{dist}_{\partial(\Omega_{\delta_k})}}{2\delta_k} \in W^{1,\infty}(U)\,.$$

We can suppose that $\delta_k \le \frac{1}{k}$, we readily get

$$\chi_k^{\mathrm{sym}} \to \tfrac{1}{2}\chi_{\partial\Omega} + \chi_{\mathrm{int}\,\Omega} \quad \text{pointwise on } U\,,$$

and, as before,

$$\|D\chi_k^{\mathrm{sym}}\|_{\mathcal{L}^1(U)} = \fint_{(-\delta_k,\delta_k)} \mathcal{H}^{n-1}(\partial\Omega_\delta)\, d\delta\,.$$

Thus we receive a good approximation for χ_Ω and a related symmetric normal measure by

$$\chi^{\mathrm{sym}} := \tfrac{1}{2}\chi_{\partial\Omega} + \chi_{\mathrm{int}\,\Omega} \quad \text{and} \quad \nu^{\mathrm{sym}} \in \mathrm{ba}\big(U, \mathcal{B}(U), \mathcal{L}^n\big)^n\,.$$

Obviously each $A = (\partial\Omega)_{\delta_k}$ is an aura of ν^{sym} by (4.32). We can interpret the factor $\frac{1}{2}$ in this way that one half of the source in a boundary point is considered to flow outward while the other half flows inward.

Now we assume that $\Omega \Subset U$ is a set of finite perimeter and consider

$$\tilde{\chi}_k^{\mathrm{sym}} := \psi_k$$

with ψ_k as in (4.42). Then, by (4.43),

$$\tilde{\chi}_k^{\mathrm{sym}} \to \tfrac{1}{2}\chi_{\partial_*\Omega} + \chi_{\mathrm{int}_*\Omega} \quad \mathcal{H}^{n-1}\text{-a.e. on } U.$$

With (4.45) for $B = U$ we see that we obtain a good approximation for χ_Ω and a related symmetric normal measure by

$$\tilde{\chi}^{\mathrm{sym}} := \tfrac{1}{2}\chi_{\partial_*\Omega} + \chi_{\mathrm{int}_*\Omega} \quad \text{and} \quad \tilde{\nu}^{\mathrm{sym}} \in \mathrm{ba}\big(U, \mathcal{B}(U), \mathcal{L}^n\big)^n\,.$$

Theorem 4.19 and (4.45) imply that

$$|\tilde{\nu}_{\mathrm{sym}}|(U) \le \mathcal{H}^{n-1}(\partial_*\Omega)$$

and, by (4.35), one has equality if $\|\tilde{\chi}^{\mathrm{sym}} - \chi_\Omega\|_{\mathcal{L}^1(U)} = 0$.

Remark 4.24

(1) While the normal measures based on a distance function can be represented by means of a normal field and a scalar density measure (cf. Proposition 4.30 below), the normal measures based on mollifications have the advantage that they are available for all sets of finite perimeter. Instead of a distance function one can also construct normal measures based on other Lipschitz continuous functions vanishing on the boundary $\partial\Omega$ (cf. [43, p. 449]).
(2) We have that (4.49), (4.52) and (4.53) are satisfied if Ω has finite perimeter and if it satisfies (4.12) (cf. (4.13) and [29]). This is in particular the case if Ω has Lipschitz boundary (cf. also the arguments following Proposition 4.7).

Clearly, each $F \in \mathcal{L}^\infty(\Omega, \mathbb{R}^n)$ is ν-integrable for any normal measure ν. Let us analyze to what extent also unbounded vector fields are integrable.

Proposition 4.25 *Let us assume that $U \subset \mathbb{R}^n$ is open and bounded, let $\Omega \in \mathcal{B}(U)$, let $F \in \mathcal{L}^1(\Omega, \mathbb{R}^n)$, let ν be a normal measure related to a good approximation χ for χ_Ω and with approximating sequence $\{\chi_k\}$ satisfying $|D\chi_k| \le \gamma k\, \mathcal{L}^n$-a.e. on Ω for some $\gamma > 0$, and let $A \subset U$ be an aura of ν as in (4.32). If there is some $\tilde{\delta} > 0$ such that*

$$\frac{1}{\delta}\int_{(\partial\Omega)_\delta \cap A} |F|\, d\mathcal{L}^n \quad \textit{is uniformly bounded for} \quad 0 < \delta < \tilde{\delta}\,, \tag{4.54}$$

then φF is ν-integrable on U for all $\varphi \in \mathcal{L}^\infty(U)$. If, in addition,

$$\lim_{k\to\infty} \frac{1}{\delta}\int_{(\partial\Omega)_\delta \cap A \cap \{|F|\ge k\}} |F|\, d\mathcal{L}^n = 0 \quad \textit{uniformly for } \delta \in (0, \tilde{\delta})\,, \tag{4.55}$$

then for each $\varphi \in \mathcal{L}^\infty(U)$ there is a subsequence $\{\chi_{k'}\}$ such that

$$\lim_{k'\to\infty} \int_U \varphi F \cdot D\chi_{k'}\, d\mathcal{L}^n = -\fint_{\partial\Omega} \varphi F\, d\nu\,. \tag{4.56}$$

Example 4.26 below shows that (4.54) is not sufficient for (4.55) and that (4.55) excludes certain concentrations on the boundary $\partial\Omega$.

Proof We argue similar to the proof of Proposition 2.15 and use

$$F = (F^1, \dots, F^n)\,, \quad \nu = (\nu^1, \dots, \nu^n)\,.$$

By assumption there is some $c > 0$ such that $\frac{1}{\delta}\int_{(\partial\Omega)_\delta \cap A} |F|\, d\mathcal{L}^n \le c$ for all $\delta \in (0, \tilde{\delta})$. For each $k \in \mathbb{N}$ and $j = 1, \dots, n$ we set

$$A_k^j := \{y \in A \mid |F^j(y)| < k\}\,, \quad A_k^{j,0} := A \setminus A_k^j\,,$$

$$h_k^j(x) := \begin{cases} \frac{l}{2^k} & \text{on } |F^j|^{-1}\big(\big[\frac{l}{2^k}, \frac{l+1}{2^k}\big)\big) \cap A_k^j \text{ for all } l \in \mathbb{N}\,, \\ 0 & \text{on } A_k^{j,0}\,. \end{cases}$$

Then all h_k^j are simple functions related to ν^j and $h_k^j \le |F^j|$ on A for all j and all $k \in \mathbb{N}$. Using $|\,h_k^j - |F^j|\,| < \frac{1}{2^k}$ on A_k^j we get

$$\big\{y \in A \;\big|\; |\,h_k^j - |F^j|\,| > \varepsilon\big\} \subset A_k^{j,0} \quad \text{if} \quad \tfrac{1}{2^k} < \varepsilon\,.$$

Therefore, for any $\varepsilon > 0$ and all j,

$$\limsup_{k\to\infty} |\nu^j|\big(\{|\,h_k^j - |F^j|\,| > \varepsilon\}\big) \le \limsup_{k\to\infty} |\nu^j|(A_k^{j,0})\,. \tag{4.57}$$

For fixed j, k there is $\tilde{\varphi} \in \mathcal{L}^\infty(U, \mathbb{R}^n)$ with $\|\tilde{\varphi}\|_\infty \le 1$ and $\tilde{\varphi} = 0$ on A_k^j such that

$$\tfrac{1}{2}|\nu^j|(A_k^{j,0}) \le \int_{A_k^{j,0}} \tilde{\varphi}\, d\nu\,.$$

(cf. (2.9)). By (4.33) there is a subsequence $\{\chi_{l'}\}$ with

$$\int_{A_k^{j,0}} \tilde{\varphi}\, d\nu = -\lim_{l'\to\infty} \int_{A_k^{j,0}} \tilde{\varphi} \cdot D\chi_{l'}\, d\mathcal{L}^n \le \liminf_{l'\to\infty} \int_{A_k^{j,0}} |D\chi_{l'}|\, d\mathcal{L}^n\,.$$

Setting $\delta_{l'} := \frac{2}{l'}$ we have $|D\chi_{l'}| \le \gamma l' = \frac{2\gamma}{\delta_{l'}}$. Thus, for l' large,

$$\begin{aligned} c &\ge \frac{1}{\delta_{l'}} \int_{(\partial\Omega)_{\delta_{l'}} \cap A} |F^j|\, d\mathcal{L}^n \\ &= \frac{1}{\delta_{l'}} \Big(\int_{(\partial\Omega)_{\delta_{l'}} \cap A_k^j} |F^j|\, d\mathcal{L}^n + \int_{(\partial\Omega)_{\delta_{l'}} \cap A_k^{j,0}} |F^j|\, d\mathcal{L}^n \Big) \\ &\ge \frac{k}{2\gamma} \int_{(\partial\Omega)_{\delta_{l'}} \cap A_k^{j,0}} |D\chi_{l'}|\, d\mathcal{L}^n = \frac{k}{2\gamma} \int_{A_k^{j,0}} |D\chi_{l'}|\, d\mathcal{L}^n \end{aligned}$$

(recall that $D\chi_{l'} = 0$ $\mathcal{L}^n$-a.e. outside $(\partial\Omega)_{\delta_{l'}}$, cf. [22, p. 130]). Taking the limit $l' \to \infty$ and using the previous estimates, we obtain

$$\frac{4c\gamma}{k} \ge |\nu^j|(A_k^{j,0}) \quad \text{for all } j \text{ and all } k \in \mathbb{N}\,.$$

Thus $|\nu^j|(A_k^{j,0}) \to 0$ as $k \to \infty$ and, using (4.57), we get

$$h_k^j \xrightarrow{\nu^j} |F^j| .$$

For the measure $h_k^j|\nu^j|$ there is some $\tilde{\varphi} \in \mathcal{L}^\infty(U)$ with $\|\tilde{\varphi}\|_\infty \le 1$ such that

$$\int_A h_k^j \, d|\nu^j| \le \int_A \tilde{\varphi} h_k^j \, d\nu^j + 1$$

(cf. (2.9)) and we use $\varphi_k \in \mathcal{L}^\infty(U, \mathbb{R}^n)$ with

$$\varphi_k = \begin{cases} \big(0, \dots, 0, \tilde{\varphi} h_k^j, 0, \dots, 0\big) & \text{on } A\,, \\ 0 & \text{on } U \setminus A\,. \end{cases}$$

By (4.33) there is some $m \in \mathbb{N}$ (related to φ_k) such that we have, with $\delta_m = \frac{2}{m}$, that $|D\chi_m| \le \gamma m$, $|\varphi_k| \le h_k^j \le |F|$, and

$$\begin{aligned} \int_A h_k^j \, d|\nu^j| &\le \int_A \tilde{\varphi} h_k^j \, d\nu^j + 1 \\ &= \int_U \varphi_k \, d\nu + 1 \;\le\; \left| \int_U \varphi_k \cdot D\chi_m \, d\mathcal{L}^n \right| + 2 \\ &\le \int_U |\varphi_k| \, |D\chi_m| \, d\mathcal{L}^n + 2 \;\le\; \gamma m \int_{(\partial\Omega)_{\delta_m} \cap A} |\varphi_k| \, d\mathcal{L}^n + 2 \\ &\le \frac{2\gamma}{\delta_m} \int_{(\partial\Omega)_{\delta_m} \cap A} |F| \, d\mathcal{L}^n + 2 \;\le\; 2\gamma c + 2 \end{aligned}$$

(notice that $\operatorname{supp} D\chi_m \subset (\partial\Omega)_{\delta_m}$). Therefore the integrals on the left hand side are uniformly bounded. Since, by construction, the sequence $\{h_k^j\}_k$ of simple functions is increasing,

$$\int_A |h_k^j - h_l^j| \, d|\nu^j| \to 0 \quad \text{as} \quad k, l \to \infty\,.$$

Consequently, $|F^j|$ is ν^j-integrable with determining sequence $\{h_k^j\}_k$ and thus, also F^j is ν^j-integrable. But this means that F is ν-integrable.

Any $\varphi \in \mathcal{L}^\infty(U)$ is ν-integrable and hence also ν-measurable. Thus also φF is ν-measurable (cf. [4, p. 102]). From $|\varphi F| \le \|\varphi\|_\infty |F|$ $\mathcal{L}^n$-a.e., we get that inequality also ν-a.e. (since $\nu <<^w \mathcal{L}^n$). Consequently, we have this estimate also i.m. ν. Hence φF is ν-integrable too (cf. [4, p. 113]).

For the second assertion we consider on U

$$F_k^j(x) := \begin{cases} F^j(x) & \text{if } |F^j(y)| < k\,, \\ 0 & \text{otherwise}\,. \end{cases} \tag{4.58}$$

Then we get as in the first part that $F_k^j \overset{\nu^j}{\to} F^j$ for each j. For $\varphi \in \mathcal{L}^\infty(U)$ we have that φF is ν-integrable and, as above, $|\varphi F_k^j| \le |\varphi F^j|$ i.m. ν. Thus, dominated convergence gives with $F_k := (F_k^1, \dots, F_k^n)$

$$\lim_{k\to\infty} \fint_{\partial\Omega} \varphi F_k \, d\nu = \fint_{\partial\Omega} \varphi F \, d\nu\,.$$

Hence, for given $\varepsilon > 0$, there is some $k_0 \in \mathbb{N}$ such that

$$\left| \fint_{\partial\Omega} \varphi F_k \, d\nu - \fint_{\partial\Omega} \varphi F \, d\nu \right| \le \varepsilon \quad \text{for all } k > k_0\,. \tag{4.59}$$

Since $|F| \le \sqrt{n}\,|F|_\infty$, we have for subsets of U that

$$\tilde{U}_k := \big\{|F|_\infty \ge k\big\} \subset \big\{|F| \ge k\sqrt{n}\big\} \subset U_k := \big\{|F| \ge k\big\}.$$

Then, by construction, $F_k = F$ on $U \setminus \tilde{U}_k$. We now choose some $l_0 \in \mathbb{N}$ with

$$\operatorname{supp}(D\chi_l) \subset A \quad \text{and} \quad \tfrac{2}{l} < \tilde{\delta} \quad \text{for all} \quad l \ge l_0$$

(cf. (4.32)). By $\operatorname{supp}(D\chi_l) \subset (\partial\Omega)_{\frac{1}{l}}$ and $|D\chi_l| \le \gamma l$ we get with some possibly larger $k_0 \in \mathbb{N}$,

$$\begin{aligned} \left| \int_U \varphi(F_k - F) \cdot D\chi_l \, d\mathcal{L}^n \right| &\le \|\varphi\|_\infty \gamma l \int_{(\partial\Omega)_{2/l} \cap A} |F_k - F| \, d\mathcal{L}^n \\ &= \|\varphi\|_\infty \gamma l \int_{(\partial\Omega)_{2/l} \cap A \cap \tilde{U}_k} |F_k - F| \, d\mathcal{L}^n \\ &\overset{|F_k| \le |F|}{\le} 2\|\varphi\|_\infty \gamma l \int_{(\partial\Omega)_{2/l} \cap A \cap U_k} |F| \, d\mathcal{L}^n \\ &\overset{(4.55)}{<} \varepsilon \end{aligned}$$

for all $k > k_0$ and $l > l_0$. Hence

$$\int_U \varphi F_k \cdot D\chi_l \, d\mathcal{L}^n - \varepsilon \le \int_U \varphi F \cdot D\chi_l \, d\mathcal{L}^n \le \int_U \varphi F_k \cdot D\chi_l \, d\mathcal{L}^n + \varepsilon\,.$$

Let us now fix $k > k_0$. Then, by (4.33) and $\varphi F_k \in \mathcal{L}^\infty(U, \mathbb{R}^n)$, there is a subsequence $\{\chi_{l'}\}$ with

$$\lim_{l' \to \infty} \int_U \varphi F_k \cdot D\chi_{l'} \, d\mathcal{L}^n = -\fint_{\partial\Omega} \varphi F_k \, d\nu \, .$$

Therefore,

$$-\fint_{\partial\Omega} \varphi F_k \, d\nu - \varepsilon \le \liminf_{l' \to \infty} \int_U \varphi F \cdot D\chi_{l'} \, d\mathcal{L}^n \le -\fint_{\partial\Omega} \varphi F_k \, d\nu + \varepsilon \, .$$

By (4.59)

$$-\fint_{\partial\Omega} \varphi F \, d\nu - 2\varepsilon \le \liminf_{l' \to \infty} \int_U \varphi F \cdot D\chi_{l'} \, d\mathcal{L}^n \le -\fint_{\partial\Omega} \varphi F \, d\nu + 2\varepsilon \, .$$

We obviously get the same estimate with limsup. Then the arbitrariness of $\varepsilon > 0$ gives the assertion. □

Though the measure of the set where an integrable function is large has to be small, the next example shows that (4.54) is not sufficient for (4.55) and (4.56).

Example 4.26 Let $U = \Omega = (0, 2)^2 \subset \mathbb{R}^2$ and let $F = (0, f)$ with

$$f(x, y) := \begin{cases} \frac{1}{y} & \text{for } 1 < x < 1 + y \, , \; y < \frac{1}{2} \, , \\ 0 & \text{otherwise} \, . \end{cases}$$

Furthermore, let ν be any normal measure as in Proposition 4.25. Since

$$\int_0^2 f(x, y) \, dx = \int_1^{1+y} \tfrac{1}{y} \, dx = 1 \quad \text{for all } \; y \in (0, 1) \, ,$$

Fubini's theorem implies $F \in \mathcal{L}^1(\Omega, \mathbb{R}^2)$ and (4.54) with $A = \Omega$. Then F is ν-integrable. Moreover, for $k \in \mathbb{N}$ and $0 < \delta < \frac{1}{k}$,

$$\frac{1}{\delta} \int_{(\partial\Omega)_\delta \cap \Omega \cap \{|F| \ge k\}} |F| \, d\mathcal{L}^2 = \frac{1}{\delta} \int_0^\delta \int_1^{1+y} f(x, y) \, dx dy = 1$$

and thus, (4.55) is not satisfied. To check (4.56) we consider an approximating sequence $F_k = (F_{1k}, F_{2k})$ of F as in (4.58). Obviously $F_k = 0$ on a small neighborhood of $\partial\Omega$. Then, for the normal measure ν of Ω and any $\varphi \in \mathcal{L}^\infty(\Omega)$, we can use $\operatorname{core} \nu \subset \partial\Omega$ and dominated convergence to get

$$0 = \fint_{\partial\Omega} \varphi F_k \, d\nu \to \fint_{\partial\Omega} \varphi F \, d\nu = 0 \, .$$

But taking e.g. χ_k from (4.50) with $\delta_k = \frac{1}{k}$, that is related to the interior normal measure ν^{int}, we readily obtain

$$\lim_{k\to\infty} \int_{\Omega} F \cdot D\chi_k \, d\mathcal{L}^2 = \lim_{k\to\infty} k \int_0^{\frac{1}{k}} \int_1^{1+y} f(x,y)\, dxdy = 1\,.$$

Hence (4.56) is not satisfied for $\varphi \equiv 1$.

Now we show how normal measures can be used for Gauss-Green formulas where we even allow some weight on the boundary $\partial\Omega$.

Theorem 4.27 *Let $U \subset \mathbb{R}^n$ be open and bounded, let $\Omega \in \mathcal{B}(U)$, let ν be a normal measure of Ω related to a good approximation χ for χ_Ω with approximating sequence $\{\chi_k\}$, let $F \in \mathcal{DM}^1(U)$ be ν-integrable such that* (4.56) *is satisfied, and let $\chi_k \to \chi$ $\operatorname{div} F$-a.e. on $\partial\Omega$. Then we have for all functions $\varphi \in \mathcal{W}^{1,\infty}(U)$ that*

$$\int_{\partial\Omega} \chi \, d \operatorname{div}(\varphi F) + \operatorname{div}(\varphi F)(\operatorname{int}\Omega) = ⨍_{\partial\Omega} \varphi F \, d\nu\,. \tag{4.60}$$

Proof Let $\varphi \in \mathcal{W}^{1,\infty}(U)$. Then $\varphi F \in \mathcal{DM}^1(U)$ with

$$\operatorname{div}(\varphi F) = \varphi \operatorname{div} F + F \cdot D\varphi \mathcal{L}^n \tag{4.61}$$

as measures on U (cf. [43, p. 448]). By the assumption and by the convergence $\chi_k \to \chi$ $\mathcal{H}^{n-1}$-a.e. on U, we get $\chi_k \to \chi$ $\operatorname{div}(\varphi F)$-a.e. on $\partial\Omega$. By $\chi_k \to \chi = 1$ everywhere on $\operatorname{int}\Omega$ and by $\chi = 0$ on $\operatorname{ext}\Omega$, dominated convergence gives

$$\int_{\partial\Omega} \chi \, d \operatorname{div}(\varphi F) + \operatorname{div}(\varphi F)(\operatorname{int}\Omega) = \int_U \chi \, d \operatorname{div}(\varphi F) = \lim_{k\to\infty} \int_U \chi_k \, d \operatorname{div}(\varphi F)\,.$$

Since $\chi_k \in W^{1,\infty}(U)$ is compactly supported on U, the definition of divergence measure (cf. (3.14)) and (4.56) with $\chi_{k'}$ related to φ imply

$$\lim_{k'\to\infty} \int_U \chi_{k'} \, d \operatorname{div}(\varphi F) = - \lim_{k'\to\infty} \int_U \varphi F \cdot D\chi_{k'} \, d\mathcal{L}^n = ⨍_{\partial\Omega} \varphi F \, d\nu$$

which implies (4.60). □

Remark 4.28

(1) For $\chi = \chi_\Omega$ the left hand side in (4.60) just equals $\operatorname{div}(\varphi F)(\Omega)$ and, in this case, we can choose

$$\lambda_F = F\nu\,, \quad \mu_F = 0$$

in Theorem 4.1 (cf. also Proposition 4.5). Notice that we can ensure ν-integrability of F by (4.54) and, due to

$$\int_{\Omega} \left| F D\chi_{\delta}^{\partial\Omega} \right| d\mathcal{L}^n \le \frac{2}{\delta} \int_{\Omega_\delta} |F|\, d\mathcal{L}^n ,$$

(4.54) also implies (4.10). Thus, for $\chi = \chi_\Omega$, we get $\mu_F = 0$ and core $\lambda_F \subset \partial\Omega$ from (4.54) already through Propositions 4.7 without using (4.56). However we do not obtain this way that $\lambda_F = F\nu$ with some normal measure ν.

(2) For any good approximations χ_1, χ_2 for χ_Ω and associated normal measures ν_1, ν_2 where $\chi_{j,k} \to \chi_j$ div F-a.e. on $\partial\Omega$, Theorem 4.27 implies

$$\int_{\partial\Omega} (\chi_1 - \chi_2)\, d\,\mathrm{div}(\varphi F) = ⨍_{\partial\Omega} \varphi F\, d(\nu_1 - \nu_2)$$

for all $\varphi \in \mathcal{W}^{1,\infty}(U)$. By (4.61) we can interchange ν_1 and ν_2 in (4.60) as long as $\chi_1 = \chi_2$ $\mathcal{L}^n$-a.e. and div F-a.e. on $\partial\Omega$. This in particular implies that $⨍_{\partial\Omega} \varphi F\, d\nu$ is independent of the choice of the good approximation χ and the corresponding normal measure ν if $|\mathrm{div}\, F|(\partial\Omega) = \mathcal{L}^n(\partial\Omega) = 0$.

(3) Since $\chi = 1$ on int Ω and

$$\big(\chi\, \mathrm{div}(\varphi F)\big)(\Omega) = \int_{\partial\Omega} \chi\, d\,\mathrm{div}(\varphi F) + \mathrm{div}(\varphi F)(\mathrm{int}\,\Omega),$$

we readily see from (4.60) that $\varphi \to \big(\chi\, \mathrm{div}(\varphi F)\big)(\Omega)$ is a trace on $\partial\Omega$ over $\mathcal{W}^{1,\infty}(U)$ under the assumptions of Theorem 4.27.

For $F \in \mathcal{DM}^\infty(U)$ we trivially have that F is ν-integrable for any normal measure ν and, by (4.33) in Theorem 4.19, we have (4.56) without the assumptions of Proposition 4.25. Since div $F <\!\!<^{w} \mathcal{H}^{n-1}$ in this case (cf. [41, p. 21]), we also have $\chi_k \to \chi$ div F-a.e. on $\partial\Omega$ by the definition of a good approximation. Thus Theorem 4.27 and Theorem 4.19 directly imply the next special case.

Corollary 4.29 *Let $U \subset \mathbb{R}^n$ be open and bounded, let $\Omega \in \mathcal{B}(U)$, let ν be a normal measure of Ω related to a good approximation χ for χ_Ω, and let $F \in \mathcal{DM}^\infty(U)$. Then F is ν-integrable, it satisfies* (4.56), *and we have* (4.60) *for all $\varphi \in \mathcal{W}^{1,\infty}(U)$. If F is even continuous and $\|\chi - \chi_\Omega\|_{\mathcal{L}^1(U)} = 0$, then*

$$\int_{\partial\Omega} \chi\, d\,\mathrm{div}(\varphi F) + \mathrm{div}(\varphi F)(\mathrm{int}\,\Omega) = \int_{\partial_*\Omega\cap U} \varphi F \cdot \nu^\Omega\, d\mathcal{H}^{n-1}$$

for all $\varphi \in C_c^1(U)$.

Let us mention that it would be possible in this case to consider ν as trace on $\partial\Omega$ over $\mathcal{DM}^\infty(U)$. This way one would not need a trace over $\mathcal{W}^{1,\infty}(U)$ for each F. But both strategies give essentially the same result. For the representation of such a trace one can use that $\mathcal{DM}^\infty(U) \subset \mathcal{L}^\infty(U)^n$ and that $\varphi F \in \mathcal{DM}^\infty(U)$ for all $\varphi \in \mathcal{BV}(U) \cap \mathcal{L}^\infty(U)$ and $F \in \mathcal{DM}^\infty(U)$ (cf. [9, p. 1014], [35, p. 65]). We now derive a more explicit structure for the normal measures

$$\nu^{\mathrm{int}}, \quad \nu^{\mathrm{intc}}, \quad \nu^{\mathrm{ext}}, \quad \text{and} \quad \nu^{\mathrm{sym}}$$

where we use the normal field ν^Ω introduced in (4.2).

Proposition 4.30 *Let $U \subset \mathbb{R}^n$ be open and bounded, let $\Omega \in \mathcal{B}(U)$, let ν^* be a normal measure where $*$ stands for* int, intc, ext, *or* sym *and let χ^*, χ_k^*, and δ_k be related to ν^* as in the corresponding examples above. Then there is some measure $\omega_{\partial\Omega}^* \in \mathrm{ba}\big(U, \mathcal{B}(U), \mathcal{L}^n\big)$ with $\operatorname{core} \omega_{\partial\Omega}^* \subset \partial\Omega$ such that*

$$\nu^* = \nu^\Omega \omega_{\partial\Omega}^* \tag{4.62}$$

and for any $\varphi \in \mathcal{L}^\infty(U)$ there is a subsequence $\chi_{k'}^*$ with

$$\int_U \varphi \, d\omega_{\partial\Omega}^* = \lim_{k' \to \infty} \frac{1}{\delta_{k'}} \int_{(\partial\Omega)_{\delta_{k'}}} \varphi \psi^* \, d\mathcal{L}^n \qquad \textit{where} \tag{4.63}$$

$$\psi^{\mathrm{int}} := \chi_{\mathrm{int}\,\Omega}\,, \quad \psi^{\mathrm{intc}} := \psi_k^{\mathrm{intc}} = 2\chi_{\Omega_{-\delta_k/2}}\,, \quad \psi^{\mathrm{ext}} := \chi_{\mathrm{ext}\,\Omega}\,, \quad \psi^{\mathrm{sym}} := \tfrac{1}{2}\,.$$

If, in addition, $F \in \mathcal{DM}^1(U)$ is ν^*-integrable such that (4.56) *is satisfied and that $\chi_k^* \to \chi^*$ $\operatorname{div} F$-a.e. on $\partial\Omega$, then we have for all $\varphi \in \mathcal{W}^{1,\infty}(U)$*

$$\operatorname{div}(\varphi F)(\operatorname{int}\Omega) = \fint_{\partial\Omega} \varphi F \cdot \nu^\Omega \, d\omega_{\partial\Omega}^{\mathrm{int}}\,,$$

$$\operatorname{div}(\varphi F)(\operatorname{int}\Omega) = \fint_{\partial\Omega} \varphi F \cdot \nu^\Omega \, d\omega_{\partial\Omega}^{\mathrm{intc}}\,,$$

$$\operatorname{div}(\varphi F)(\overline{\Omega}) = \fint_{\partial\Omega} \varphi F \cdot \nu^\Omega \, d\omega_{\partial\Omega}^{\mathrm{ext}}\,,$$

$$\tfrac{1}{2}\operatorname{div}(\varphi F)(\partial\Omega) + \operatorname{div}(\varphi F)(\operatorname{int}\Omega) = \fint_{\partial\Omega} \varphi F \cdot \nu^\Omega \, d\omega_{\partial\Omega}^{\mathrm{sym}}\,.$$

Notice that the measures $\omega_{\partial\Omega}^*$, that can be considered as some surface area, are measures of the type constructed in Proposition 3.1. In the case ν^{int} we have to take $\gamma(\delta) = \delta$ and $E = \operatorname{int}\Omega$ there. However, here we cannot just apply Proposition 3.1, since we have to select that weak* cluster point of the measures $\frac{1}{\delta_k}\chi_{(\partial\Omega)_{\delta_k} \cap \operatorname{int}\Omega}\mathcal{L}^n$ that is related to ν^{int}. In the proof we see that the subsequence $\chi_{k'}^*$ in (4.63) is the same as that for $\varphi\nu^\Omega$ in (4.33). We refer to the arguments following (4.2) for the properties of the normal field ν^Ω. Notice that the previous proposition is applicable to all $F \in \mathcal{DM}^\infty(U)$, since (4.56) is always satisfied in this case (cf. (4.33) in Theorem 4.19 or Corollary 4.29) and $\operatorname{div} F <<^w \mathcal{H}^{n-1}$ (cf. [41, p. 21]).

The explicit occurrence of the normal field ν^Ω in the Gauss-Green formulas above is due to the fact that, for these normal measures, the normalized gradient of the associated $D\chi_k$ equals the gradient of the distance function on the support of $D\chi_k$ and, thus, it is independent of k near $\partial\Omega$. However, this is not met for the normal measures based on mollification. Therefore we cannot go beyond (4.60) in those cases in general. Let us also refer to the fact that the Radon-Nikodym theorem is only available "up to a small error $\varepsilon > 0$" for normal measures ν that are typically pure (cf. [4, p. 191]).

Proof Let us first consider ν^{int} with $\chi_k = \chi_k^{\mathrm{int}}$, δ_k as in (4.49) and (4.50). Then

$$D\chi_k = \begin{cases} -\frac{1}{\delta_k}\nu^\Omega & \mathcal{L}^n\text{-a.e. on } \operatorname{int}\Omega \setminus \overline{\Omega_{-\delta_k}}\,, \\ 0 & \text{otherwise}\,. \end{cases} \tag{4.64}$$

Obviously

$$\varphi \to \int_U \varphi\nu^\Omega \, d\nu^{\mathrm{int}} \quad \text{for} \quad \varphi \in \mathcal{L}^\infty(U)$$

belongs to $\mathcal{L}^\infty(U)^*$ and can be identified with some measure

$$\omega_{\partial\Omega}^{\mathrm{int}} \in \mathrm{ba}\big(U, \mathcal{B}(U), \mathcal{L}^n\big)$$

(cf. [4, p. 106]). Then

$$\int_U \varphi \, d\omega_{\partial\Omega}^{\mathrm{int}} = \int_U \varphi\nu^\Omega \, d\nu^{\mathrm{int}} \quad \text{for all} \quad \varphi \in \mathcal{L}^\infty(U)\,.$$

For $\Phi \in \mathcal{L}^\infty(U)^n$ there is a pointwise unique orthogonal decomposition $\mathcal{L}^n$-a.e. such that

$$\Phi = (\Phi \cdot \nu^\Omega)\,\nu^\Omega + \Phi^\perp \quad \text{where} \quad \Phi^\perp \cdot \nu^\Omega = 0\,.$$

Then $⨍_{\partial\Omega} \Phi^\perp \, d\nu^{\mathrm{int}} = 0$ by (4.33) and (4.64). Thus, for all $\Phi \in \mathcal{L}^\infty(U)^n$,

$$⨍_{\partial\Omega} \Phi \, d\nu^{\mathrm{int}} = ⨍_{\partial\Omega} (\Phi \cdot \nu^\Omega)\,\nu^\Omega \, d\nu^{\mathrm{int}} = ⨍_{\partial\Omega} \Phi \cdot \nu^\Omega \, d\omega_{\partial\Omega}^{\mathrm{int}}\,.$$

Consequently,

$$\nu^{\mathrm{int}} = \nu^\Omega \omega_{\partial\Omega}^{\mathrm{int}}\,.$$

For $\varphi \in \mathcal{L}^\infty(U)$ we use (4.64) and (4.33) with $\varphi\nu^\Omega$ and the corresponding subsequence $\chi_{k'}$ to get

$$\begin{aligned}\int_U \varphi\, d\omega_{\partial\Omega}^{\text{int}} = \fint_{\partial\Omega} \varphi\nu^\Omega\, d\nu^{\text{int}} &= -\lim_{k'\to\infty} \int_U \varphi\nu^\Omega \cdot D\chi_{k'}\, d\mathcal{L}^n \\ &= \lim_{k'\to\infty} \frac{1}{\delta_{k'}} \int_{(\partial\Omega)_{\delta_{k'}}} \varphi\chi_{\text{int}\,\Omega}\, d\mathcal{L}^n\,.\end{aligned}$$

For the other cases we argue analogously.

Let now $F \in \mathcal{DM}^1(U)$ be ν^*-integrable such that (4.56) is satisfied. Then the stated Gauss-Green formulas follow directly from Proposition 4.27 and (4.62) with the related χ^*. □

The next example shows how the different cases in the previous proposition work.

Example 4.31 Let $U = B_3(0) \subset \mathbb{R}^2$, let $\Omega = \Omega_1 \cup \Omega_2$ with

$$\Omega_1 = (0,1)^2\,, \quad \Omega_2 = (1,2)\times(0,1)\,,$$

and consider

$$F(x,y) = \begin{cases} (1,0) & \text{on } \Omega_1\,, \\ (2,0) & \text{for } x > 1\,, \\ (0,0) & \text{otherwise.}\end{cases}$$

For χ_k related to ν^{ext} and $\delta > 0$ small we obviously have that

$$D\chi_k = 0 \quad \text{on} \quad (1-\delta, 1+\delta)\times(0,1)\,.$$

Then, by Proposition 4.30 with $\varphi \equiv 1$, by (4.56), and for $\delta > 0$ small, we readily get

$$\begin{aligned}\operatorname{div} F(\Omega) = \fint_{\partial\Omega} F\cdot\nu^\Omega\, d\omega_{\partial\Omega}^{\text{int}} &= \fint_{\partial\Omega} F\, d\nu^{\text{int}} \\ &= (F\nu^{\text{int}})\big(\{x\in(0,\delta)\}\big) + (F\nu^{\text{int}})\big(\{x\in(1-\delta,1+\delta)\}\big) \\ &\quad + (F\nu^{\text{int}})\big(\{x\in(2-\delta,2)\}\big) \\ &= -1 + (1-2) + 2 = 0\,,\end{aligned}$$

$$\begin{aligned}\operatorname{div} F(\overline{\Omega}) = \fint_{\partial\Omega} F\cdot\nu^\Omega\, d\omega_{\partial\Omega}^{\text{ext}} &= \fint_{\partial\Omega} F\, d\nu^{\text{ext}} \\ &= (F\nu^{\text{ext}})\big(\{x\in(-\delta,0)\}\big) + (F\nu^{\text{ext}})\big(\{x\in(1-\delta,1+\delta)\}\big)\end{aligned}$$

$$+(F\nu^{\mathrm{ext}})\big(\{x \in (2, 2+\delta)\}\big)$$
$$= 0 + 0 + 2 = 2,$$

$\frac{1}{2}\mathrm{div}\, F(\partial\Omega) + \mathrm{div}\, F(\mathrm{int}\,\Omega)$

$$= \fint_{\partial\Omega} F \cdot \nu^{\Omega}\, d\omega_{\partial\Omega}^{\mathrm{sym}} = \fint_{\partial\Omega} F\, d\nu^{\mathrm{sym}}$$
$$= (F\nu^{\mathrm{sym}})\big(\{x \in (-\delta, \delta)\}\big) + (F\nu^{\mathrm{sym}})\big(\{x \in (1-\delta, 1+\delta)\}\big)$$
$$+ (F\nu^{\mathrm{sym}})\big(\{x \in (2-\delta, 2+\delta)\}\big)$$
$$= -\tfrac{1}{2} + (\tfrac{1}{2} - 1) + 2 = 1.$$

Let us now consider the application to unbounded vector fields (cf. also Chen-Frid [6, p. 403]).

Example 4.32 Let $U = B_2(0) \subset \mathbb{R}^2$, let $\Omega = (0, 1)^2$, and take $F = (F^1, F^2)$ given by

$$F(x) := \frac{x}{2\pi\, |x|^2} \quad \text{with} \quad \mathrm{div}\, F = \delta_0 \quad \big(x = (x_1, x_2)\big)$$

(cf. Example 4.12). We consider the normal measure $\nu = \nu^{\mathrm{int}} = (\nu^1, \nu^2)$ and, according to Proposition 4.30, we have

$$\nu = \nu^{\Omega} \omega_{\partial\Omega}^{\mathrm{int}}. \tag{4.65}$$

A simple computation shows that

$$\frac{1}{\delta} \int_{(\partial\Omega)_\delta \cap \Omega} |F|\, d\mathcal{L}^2 \overset{\delta\to 0}{\longrightarrow} \infty.$$

Hence we cannot use Proposition 4.25 to get ν-integrability. Therefore let us use a more direct argument. We divide Ω by its diagonals into four triangles

$$\Omega_{*0}, \;\Omega_{*1}, \;\Omega_{0*}, \;\Omega_{1*}$$

where e.g. Ω_{*0} is the triangle with a side on the line $\{x_2 = 0\}$. Clearly F is bounded and thus, ν-integrable on $\Omega_{*1} \cup \Omega_{1*}$. Using (4.33) we readily get

$$\int_{\Omega_{*1} \cup \Omega_{1*}} F\, d\nu = \frac{1}{4}.$$

On Ω_{*0} we obviously have $\nu^{\Omega} = (0, -1)$. Hence, by (4.65), ν^1 is the zero measure on Ω_{*0}. Therefore F^1 is ν^1-integrable on Ω_{*0} and the integral vanishes. Now we consider F^2 with respect to ν^2 and set for $\varepsilon > 0$

$$M_\varepsilon := \{x \in \Omega_{*0} \mid |F^2| > \varepsilon\}.$$

Using that F^2 is continuous and vanishes for $x_2 = 0$, we have for any $\kappa > 0$

$$M_\varepsilon \cap \{|x| \geq \kappa\} \cap \{x_2 = 0\}_\delta = \emptyset$$

for some small $\delta > 0$. Since $\{x_2 = 0\}_\delta$ is an aura of $\omega_{\partial\Omega}^{\text{int}}$ restricted to Ω_{*0},

$$|\nu^2|\big(M_\varepsilon \cap \{|x| \geq \kappa\}\big) = 0 \quad \text{for all } \kappa > 0.$$

Consequently, using (4.63),

$$\begin{aligned} |\nu^2|\big(\{x \in \Omega_{*0} \mid |F^2| > \varepsilon\}\big) &= |\nu^2|\big(\{x \in \Omega_{*0} \mid |F^2| > \varepsilon\,,\ |x| < \kappa\}\big) \\ &\leq |\nu^2|\big(\{x \in \Omega_{*0} \mid |x| < \kappa\}\big) \overset{(4.65)}{\leq} \kappa \quad \text{for all } \kappa > 0. \end{aligned}$$

Therefore

$$|\nu^2|\big(\{x \in \Omega_{*0} \mid |F^2| > \varepsilon\}\big) = 0.$$

This means that F^2 agrees i.m. ν^2 with the zero function on Ω_{*0}. Consequently F^2 is ν^2-integrable on Ω_{*0} and thus, the integral vanishes. Summarizing we get that

$$\int_{\Omega_{*0}} F\, d\nu = 0 \quad \text{and, analogously,} \quad \int_{\Omega_{0*}} F\, d\nu = 0.$$

Hence, though (4.54) is not satisfied, F is ν-integrable on Ω with

$$\oint_{\partial\Omega} F\, d\nu = \oint_{\partial\Omega\setminus B_1(0)} F\, d\nu = \frac{1}{4}.$$

Let us now check (4.56) directly for $\varphi \equiv 1$. With χ_k^{int} according to (4.50) we have

$$\begin{aligned} \int_{\Omega_{*0}} F \cdot D\chi_k^{\text{int}}\, d\mathcal{L}^2 &= \int_{(\partial\Omega)_{1/k}\cap\Omega_{*0}} k F^2\, d\mathcal{L}^2 \\ &= \frac{k}{2\pi} \int_0^{\frac{1}{k}} \int_{x_2}^{1-x_2} \frac{x_2}{x_1^2 + x_2^2}\, dx_1 dx_2 \\ &= \frac{k}{2\pi} \int_0^{\frac{1}{k}} x_2 \Big[\tfrac{1}{x_2} \arctan \tfrac{x_1}{x_2}\Big]_{x_1 = x_2}^{1-x_2} dx_2 \end{aligned}$$

$$= \frac{1}{2\pi} \fint_{(0,\frac{1}{k})} \left(\arctan \tfrac{1-x_2}{x_2} - \tfrac{\pi}{4} \right) dx_2$$

$$\overset{k\to\infty}{\longrightarrow} \frac{1}{2\pi} \left(\frac{\pi}{2} - \frac{\pi}{4} \right) = \frac{1}{8} .$$

By analogous arguments on the other triangles we end up with

$$\int_\Omega F \cdot D\chi_k^{\text{int}} \, d\mathcal{L}^2 \overset{k\to\infty}{\longrightarrow} 0 \neq -\frac{1}{4} = -\fint_{\partial\Omega} F \, d\nu .$$

Thus (4.56) is violated and obviously

$$(\text{div}\, F)(\Omega) = \delta_0(\Omega) = 0 \neq \fint_{\partial\Omega} F \, d\nu .$$

Therefore (4.60) does not hold and we see that this condition is essential in Theorem 4.27. As in Example 4.12 we obtain that

$$\text{div}(\varphi F)(\Omega) = \int_{\partial\Omega} F \nu^\Omega \, d\mathcal{H}^1 - \tfrac{1}{4}\varphi(0)$$

for all $\varphi \in \mathcal{W}^{1,\infty}(U)$. Since the measure $F\nu$ is absolutely continuous with respect to ν, we in fact cannot expect (4.60) with a normal measure as in (4.65) in the case of a concentration at the origin (cf. [4, p. 106]).

Let us also provide an unbounded vector field where Theorem 4.27 is applicable.

Example 4.33 For the sets $U = B_2(0) \subset \mathbb{R}^3$ and $\Omega = (0,1)^3$ we consider as in the previous example

$$F(x) := \frac{x}{2\pi |x|^2} \quad \text{where} \quad x = (x_1, x_2, x_3) .$$

Then

$$s \to \int_{(0,1)^2} |F(x_1, x_2, s)| \, d\mathcal{H}^2(x_1, x_2)$$

is continuous and bounded. Thus (4.54) is satisfied and F is integrable with respect to any normal measure ν satisfying the assumption of Proposition 4.25 (which is the case for the normal measures constructed above by means of a distance function).

For such a normal measure ν with aura in Ω we get for large $l \in \mathbb{N}$

$$\begin{aligned}
\int_{(\partial\Omega)_\delta \cap \Omega \cap \{|F| \ge l\}} & |F| \, d\mathcal{L}^3 \\
& \le \int_{B_{1/(2\pi l)}(0) \cap (\partial\Omega)_\delta \cap \Omega} |F| \, d\mathcal{L}^3 \\
& \le \frac{3}{4} \int_0^\delta \int_{B_{1/(2\pi l)}(0) \cap \{x_3 = 0\}} \frac{1}{2\pi\sqrt{x_1^2 + x_2^2}} d\mathcal{H}^2(x_1, x_2) \, dx_3 \\
& \le \frac{3}{4} \int_0^\delta \int_0^{\frac{1}{2\pi l}} \frac{2\pi r}{2\pi r} \, dr dx_3 \; = \; \frac{3\delta}{8\pi l} \, .
\end{aligned}$$

Therefore,

$$\frac{1}{\delta} \int_{(\partial\Omega)_\delta \cap \Omega \cap \{|F| \ge l\}} |F| \, d\mathcal{L}^n \le \frac{3}{8\pi l} \quad \text{for all } \delta > 0$$

which gives (4.55) and (4.56). Thus, by Proposition 4.30 e.g. for $\nu = \nu^{\text{int}}$,

$$\operatorname{div}(\varphi F)(\Omega) = \fint_{\partial\Omega} \varphi F \cdot \nu^\Omega \, d\omega_{\partial\Omega}^{\text{int}}$$

for all $\varphi \in \mathcal{W}^{1,\infty}(U)$.

Now we consider the important case of bounded vector fields $F \in \mathcal{DM}^\infty(U)$ on sets $\Omega \subset U$ with finite perimeter in some more detail. Here we use normal measures that are based on mollifications, in particular those from Examples 4.21 and 4.22. Though some of the assertions in the next proposition are already known from the literature (cf. [8, 10, 12–14, 41] and the remarks below), we include them not only for completeness but also to show their relation to the new results. In the proof, given after Remark 4.38, we essentially use arguments that are based on the theory developed here. Recall that, in addition to the subsequent results, Proposition 4.30 is applicable to Ω if $\mathcal{H}^{n-1}(\Omega_\delta)$ is bounded for $|\delta|$ small.

Proposition 4.34 *Let $U \subset \mathbb{R}^n$ be an open and bounded set, let $\Omega \subset U$ have finite perimeter, let $F \in \mathcal{DM}^\infty(U)$, and let $\tilde{\nu}^{\text{int}}$, $\tilde{\nu}^{\text{ext}}$ be the normal measures from Examples 4.21 and 4.22.*

(1) *If $\Omega \Subset U$, then there are normal trace functions*

$$f^{\text{int}}, f^{\text{ext}} \in \mathcal{L}^\infty(\partial_*\Omega, \mathcal{H}^{n-1}) \quad \textit{with} \quad \|f^{\text{int}}\|_\infty, \|f^{\text{ext}}\|_\infty \le \|F\|_{\partial\Omega}$$

such that

$$\operatorname{div}(\varphi F)(\operatorname{int}_*\Omega) = \fint_{\partial\Omega} \varphi F \, d\tilde{\nu}^{\mathrm{int}} = \int_{\partial_*\Omega} \varphi f^{\mathrm{int}} \, d\mathcal{H}^{n-1}, \tag{4.66}$$

$$\operatorname{div}(\varphi F)(\partial_*\Omega \cup \operatorname{int}_*\Omega) = \fint_{\partial\Omega} \varphi F \, d\tilde{\nu}^{\mathrm{ext}} = \int_{\partial_*\Omega} \varphi f^{\mathrm{ext}} \, d\mathcal{H}^{n-1} \tag{4.67}$$

for all $\varphi \in \mathcal{W}^{1,\infty}(U)$ *(cf.* (2.10) *for* $\|\cdot\|_{\partial\Omega}$*). If* Ω *is open we also have*

$$\operatorname{div}(\varphi F)(\Omega) = \int_{\partial_*\Omega} \varphi f^{\mathrm{int}} \, d\mathcal{H}^{n-1} - \int_{\operatorname{int}_*\Omega \cap \partial\Omega} \varphi \, d \operatorname{div} F \tag{4.68}$$

and if Ω *is closed*

$$\operatorname{div}(\varphi F)(\Omega) = \int_{\partial_*\Omega} \varphi f^{\mathrm{ext}} \, d\mathcal{H}^{n-1} + \int_{\operatorname{ext}_*\Omega \cap \partial\Omega} \varphi \, d \operatorname{div} F \tag{4.69}$$

for all $\varphi \in \mathcal{W}^{1,\infty}(U)$.

(2) *Let* $\Omega \subset U$ *be open with* $\mathcal{H}^{n-1}(\partial\Omega \cap \operatorname{int}_*\Omega) < \infty$*. Then* $\hat{F} \in \mathcal{DM}^\infty(\mathbb{R}^n)$ *for the extension* $\hat{F}$ *of* F *by zero outside of* Ω*. There is also some normal measure* $\nu \in \operatorname{ba}\big(U, \mathcal{B}(U), \mathcal{L}^n\big)^n$ *such that* $(\partial\Omega)_\delta \cap \Omega$ *is an aura for all* $\delta > 0$ *and*

$$\operatorname{div}(\varphi F)(\Omega) = \fint_{\partial\Omega} \varphi F \, d\nu \tag{4.70}$$

for all $\varphi \in \mathcal{W}^{1,\infty}(U)$*. Moreover there is* $f \in \mathcal{L}^\infty(\partial\Omega \setminus \operatorname{ext}_*\Omega, \mathcal{H}^{n-1})$ *with* $\|f\|_\infty \le c\|F\|_{\partial\Omega}$ *for some* $c > 0$ *depending only on dimension* n *such that*

$$\operatorname{div}(\varphi F)(\Omega) = \int_{\partial\Omega \setminus \operatorname{ext}_*\Omega} \varphi f \, d\mathcal{H}^{n-1} \tag{4.71}$$

for all $\varphi \in C^1(\mathbb{R}^n)$ *and*

$$f\mathcal{H}^{n-1} \llcorner (\partial\Omega \cap \operatorname{int}_*\Omega) = -\operatorname{div} F \llcorner (\partial\Omega \cap \operatorname{int}_*\Omega) \,. \tag{4.72}$$

If $\hat{F}$ *is continuous on a neighborhood of* Ω*, then* $(\operatorname{div} F)(\partial\Omega \cap \operatorname{int}_*\Omega) = 0$ *and*

$$\operatorname{div}(\varphi F)(\operatorname{int}_*\Omega) = \operatorname{div}(\varphi F)(\Omega) = \int_{\partial_*\Omega} \varphi F \cdot \nu^\Omega \, d\mathcal{H}^{n-1} \tag{4.73}$$

for all $\varphi \in C^1(\mathbb{R}^n)$.

Remark 4.35

(1) According to Proposition 2.13 the measures $f^{\text{int}}\mathcal{H}^{n-1}\lfloor\partial_*\Omega$ and $f^{\text{ext}}\mathcal{H}^{n-1}\lfloor\partial_*\Omega$ are the Radon measures related to $F\tilde{\nu}^{\text{int}}$ and $F\tilde{\nu}^{\text{ext}}$. Notice that we can only replace $\tilde{\nu}^{\text{int}}$ and $\tilde{\nu}^{\text{ext}}$ with a Radon measure in (4.66) and (4.67), respectively, if F is continuous (cf. Proposition 2.13). With the approximating sequences $\tilde{\chi}_k^{\text{int}}$, $\tilde{\chi}_k^{\text{ext}}$ corresponding to $\tilde{\nu}^{\text{int}}$, $\tilde{\nu}^{\text{ext}}$ we have that

$$\begin{aligned} &-F\cdot D\tilde{\chi}_k^{\text{int}}\mathcal{L}^n \overset{*}{\rightharpoonup} f^{\text{int}}\mathcal{H}^{n-1}\lfloor\partial_*\Omega \qquad \text{and} \\ &-F\cdot D\tilde{\chi}_k^{\text{ext}}\mathcal{L}^n \overset{*}{\rightharpoonup} f^{\text{ext}}\mathcal{H}^{n-1}\lfloor\partial_*\Omega \end{aligned} \tag{4.74}$$

as weak* limits for Radon measures. This way we basically recover several results from the literature about inner and outer traces on sets of finite perimeter (cf. e.g. Chen-Torres-Ziemer [10, pp. 275, 281], Chen-Comi-Torres [11, p. 106], Comi-Payne [13, pp. 194, 200]). Though the sequences in (4.74) slightly differ from the sequences using $\psi_k = \chi_\Omega * \eta_{\frac{1}{k}}$ and given by

$$-2\chi_\Omega F\cdot D\psi_k\mathcal{L}^n \quad \text{and} \quad -2\chi_{U\setminus\Omega}F\cdot D\psi_k\mathcal{L}^n\,,$$

that are commonly used in the literature (up to sign), the weak* limit is always the same, since in each case (4.66) and (4.67) have to be valid for a set of φ that is dense in $C_c(U)$. This in particular means that

$$\begin{aligned} \int_{\partial_*\Omega}\varphi f^{\text{int}}\,d\mathcal{H}^{n-1} &= -\lim_{k\to\infty}\int_U \varphi F\cdot D\tilde{\chi}_k^{\text{int}}\,d\mathcal{L}^n \\ &= -\lim_{k\to\infty}\int_U 2\varphi\chi_\Omega F\cdot D\psi_k\,d\mathcal{L}^n \end{aligned} \tag{4.75}$$

for all $\varphi\in C_c(U)$ and analogously for the other case.

(2) We can argue as in the proof of Proposition 4.34 below to see that the first assertion is also true for $\tilde{\nu}^{\text{sym}}$ with

$$\tfrac{1}{2}\operatorname{div}(\varphi F)(\partial_*\Omega)+\operatorname{div}(\varphi F)(\operatorname{int}_*\Omega) = \fint_{\partial\Omega}\varphi F\,d\tilde{\nu}^{\text{sym}} = \int_{\partial_*\Omega}\varphi f^{\text{sym}}\,d\mathcal{H}^{n-1}$$

for some $f^{\text{sym}}\in\mathcal{L}^\infty(\partial_*\Omega,\mathcal{H}^{n-1})$ with $\|f^{\text{sym}}\|_\infty\le\|F\|_{\partial\Omega}$. Moreover, a check of the relevant proofs also shows that we can replace the requirement $\Omega\Subset U$ in Proposition 4.34 (1) with the restriction to functions $\varphi\in\mathcal{W}^{1,\infty}(U)$ having compact support in U. In this case a further restriction to φ with

$$\varphi=0 \quad \text{on} \quad U\setminus(\partial\Omega)_\delta \quad \text{for some } \delta>0$$

is sufficient for the treatment of a Gauss-Green formula (cf. Remark 3.6). This way we recover the results from Comi-Payne [13, p. 203].

(3) The proof shows that (4.71) and (4.72) with some $f \in \mathcal{L}^1(\partial\Omega \setminus \mathrm{ext}_*\Omega, \mathcal{H}^{n-1})$ are already true without the $\mathcal{H}^{n-1}$-bound as long as the extension of F by zero has divergence measure and $\mathcal{H}^{n-1}$ is σ-finite on $\partial\Omega \setminus \mathrm{ext}_*\Omega$.

Remark 4.36 Let $U \subset \mathbb{R}^n$ be an open bounded set, let $\Omega \subset U$ be open with finite perimeter such that also $\mathrm{int}_*\Omega \subset U$, let $F \in \mathcal{DM}^1(U)$, and let σ_F^{int} be a Radon measure on $\partial_*\Omega$ with $\sigma_F^{\mathrm{int}} <<{}^w\mathcal{H}^{n-1}$ such that

$$\mathrm{div}(\varphi F)(\mathrm{int}_*\Omega) = \int_{\partial_*\Omega} \varphi \, d\sigma_F^{\mathrm{int}} \tag{4.76}$$

for all $\varphi \in \mathcal{W}^{1,\infty}(U) \cap C(\overline{U})$ (cf. (4.66)). With the disjoint decomposition

$$\mathrm{int}_*\Omega = \Omega \cup (\mathrm{int}_*\Omega \cap \partial\Omega) \tag{4.77}$$

and since Ω and $\mathrm{int}_*\Omega$ merely differ by an $\mathcal{L}^n$-null set (cf. [22, p. 222]), we easily obtain that

$$\begin{aligned}
\mathrm{div}(\varphi F)(\Omega) &= \int_\Omega F \cdot D\varphi \, d\mathcal{L}^n + \int_\Omega \varphi \, d \,\mathrm{div}\, F \\
&= \int_{\mathrm{int}_*\Omega} F \cdot D\varphi \, d\mathcal{L}^n + \int_{\mathrm{int}_*\Omega} \varphi \, d \,\mathrm{div}\, F - \int_{\mathrm{int}_*\Omega\cap\partial\Omega} \varphi \, d \,\mathrm{div}\, F \\
&= \int_{\partial_*\Omega} \varphi \, d\sigma_F^{\mathrm{int}} - \int_{\mathrm{int}_*\Omega\cap\partial\Omega} \varphi \, d \,\mathrm{div}\, F \,.
\end{aligned}$$

Since $\partial_*\Omega \subset \partial\Omega$, we use the Radon measure

$$\sigma_F := \sigma_F^{\mathrm{int}} - \mathrm{div}\, F \lfloor (\mathrm{int}_*\Omega \cap \partial\Omega) \tag{4.78}$$

to get

$$\mathrm{div}(\varphi F)(\Omega) = \int_{\partial\Omega\setminus\mathrm{ext}_*\Omega} \varphi \, d\sigma_F \tag{4.79}$$

for all $\varphi \in \mathcal{W}^{1,\infty}(U) \cap C(\overline{U})$. For $F \in \mathcal{DM}^\infty(\Omega)$ an $\mathcal{H}^{n-1}$-integrable density f of σ_F is available by the Radon-Nikodym theorem if $\mathcal{H}^{n-1}$ is σ-finite on $\partial\Omega \setminus \mathrm{ext}_*\Omega$ (cf. [2, p. 14]). This is obviously the case if $\mathcal{H}^{n-1}(\partial\Omega \setminus \mathrm{ext}_*\Omega) < \infty$ and leads to (4.71) with an essentially bounded density.

This shows that we can easily transform (4.76) to a Gauss-Green formula where the full topological boundary is incorporated into the boundary term, which is favorable in the case of cracks along some inner boundary. Notice, however, that the form with a normal measure as e.g. in (4.66) or in (4.60) contains F explicitly in the boundary term and does not require a normal trace function f on $\partial\Omega$. Moreover the larger class of functions $\mathcal{W}^{1,\infty}(U)$ for φ, that do not have to be continuous up to $\partial\Omega$, allows more flexibility for the investigation near $\partial\Omega$ (cf. the discussion surrounding (3.10)). In Example 4.13 this allows e.g., in contrast to (4.79), to describe the behavior on both sides of the inner boundary $\partial\Omega \cap \operatorname{int}_*\Omega$ by the normal components of F, which is of course helpful for the description of cracks.

In the special situation where Ω is open and bounded and where the vector field $F \in \mathcal{DM}^\infty(\Omega)$ is such that $\hat{F} \in \mathcal{DM}^\infty(\mathbb{R}^n)$ for its extension $\hat{F}$ by zero, we can choose some open bounded $\hat{U} \subset \mathbb{R}^n$ with $\Omega \Subset \hat{U}$. Applying Proposition 4.34 (1) to $\hat{F}$ we obtain (4.66) and, by the previous arguments, we get (4.79). Using also the second assertion of Proposition 4.34, we basically recover the Gauss-Green formulas stated by Chen-Li-Torres [12, p. 248]. Let us mention that a vector field $F \in \mathcal{DM}^\infty(\Omega)$ can be extended by zero as required above if Ω is a bounded open set of finite perimeter satisfying $\mathcal{H}^{n-1}(\partial\Omega \cap \operatorname{int}_*\Omega) < \infty$ (cf. [12, p. 242]).

The previous remark and the definition of divergence measure in (3.14) lead to a simple extension criterion for open Ω with finite perimeter and bounded vector fields F (cf. also [11, p. 104], [12, 13]).

Proposition 4.37 *Let $\Omega \subset \mathbb{R}^n$ be an open bounded set with finite perimeter and let $F \in \mathcal{DM}^\infty(\Omega)$. Then the extension $\hat{F}$ of F by zero belongs to $\mathcal{DM}^\infty(\mathbb{R}^n)$ if and only if there is a Radon measure σ_F supported on $\partial\Omega \setminus \operatorname{ext}_*\Omega$ such that*

$$\int_\Omega F \cdot D\varphi \, d\mathcal{L}^n + \int_\Omega \varphi \, d \operatorname{div} F = \int_{\partial\Omega \setminus \operatorname{ext}_*\Omega} \varphi \, d\sigma_F \tag{4.80}$$

for all $\varphi \in C^1(\mathbb{R}^n)$.

This basically means that the extension of F by zero has also divergence measure if and only if there is a Gauss-Green formula where the boundary term is related to a Radon measure.

Proof First we assume that the zero extension $\hat{F}$ of F belongs to $\mathcal{DM}^\infty(\mathbb{R}^n)$. Then we can argue as in Remark 4.36 to get (4.80). If otherwise (4.80) is satisfied, then we have for the zero extension $\hat{F}$ and for

$$\hat{\sigma} := (\operatorname{div} F) \lfloor \Omega - \sigma_F$$

that

$$\int_{\mathbb{R}^n} \hat{F} \cdot D\varphi \, d\mathcal{L}^n = - \int_{\mathbb{R}^n} \varphi \, d\hat{\sigma}$$

for all $C_c^1(\mathbb{R}^n)$. Thus $\hat{F} \in \mathcal{DM}^\infty(\mathbb{R}^n)$. □

Remark 4.38 Let us discuss the general case where $U \subset \mathbb{R}^n$ is open and bounded, $\Omega \in \mathcal{B}(U)$, and $F \in \mathcal{DM}^\infty(U)$. Then we always have

$$\operatorname{div}(\varphi F)(\Omega) = \int_{(\partial\Omega)_\delta \cap U} \varphi \, d\lambda_F + \int_{(\partial\Omega)_\delta \cap U} D\varphi \, d\mu_F \tag{4.81}$$

for any $\delta > 0$ and all $\varphi \in \mathcal{W}^{1,\infty}(U)$ by (4.1). If $\Omega = U$ then we have case (L) with core $\lambda_F \subset \partial\Omega$ and we can remove δ in the first integral. If $\Omega \Subset U$ then we have case (C) and we can identify λ_F with a σ-measure on $\partial\Omega$ (cf. Corollary 3.18). If merely $\Omega \in \mathcal{B}(U)$ with $\mathcal{L}^n(\Omega \setminus \operatorname{int}\Omega) = 0$ but in addition

$$\sup_{0<\delta<\tilde{\delta}} \mathcal{H}^{n-1}(\partial\Omega_{-\delta}) < \infty \quad \text{for some} \quad \tilde{\delta} > 0\,, \tag{4.82}$$

then we have

$$\operatorname{div}(\varphi F)(\Omega) = \fint_{\partial\Omega} \varphi \, d\lambda_F \tag{4.83}$$

(cf. Proposition 4.7). Notice that (4.82) is valid for a large class of sets of finite perimeter and, if $\Omega \Subset U$, it allows the application of the normal measures ν^{int}, ν^{ext}, and ν^{sym} that are constructed by means of a distance function. This leads to Gauss-Green formulas containing F and the normal field ν^Ω explicitly in the boundary term (cf. Proposition 4.30 and the subsequent discussion). If (4.82) is not available, then we can apply the normal measures $\tilde{\nu}^{\mathrm{int}}$, $\tilde{\nu}^{\mathrm{sym}}$, and $\tilde{\nu}^{\mathrm{ext}}$, that are based on mollified functions, to any set of finite perimeter $\Omega \Subset U$. Then the right hand side in (4.83) has the form $\fint_{\partial\Omega} \varphi F \, d\nu$ with ν being one of those normal measures. But here we cannot explicitly incorporate a normal field (cf. Theorem 4.27 and Corollary 4.29). We have to realize that (4.82), though it excludes some "exotic" sets of finite perimeter, leads to more structural information about λ_F in the boundary term.

Notice that, even if merely $\Omega \subset U$, it is possible in the general variants (4.81) and (4.83) to account precisely for the boundary points belonging to Ω, which is relevant if parts of the boundary belong to the support of $\operatorname{div} F$. If we want to have more structural information in a Gauss-Green formula "up to the boundary", i.e. if we e.g. have $\Omega = U$, then merely the normal measure ν^{intc} is available. It gives

$$\operatorname{div}(\varphi F)(\operatorname{int}\Omega) = \fint_{\partial\Omega} \varphi F \, d\nu^{\mathrm{intc}} = \fint_{\partial\Omega} \varphi F \cdot \nu^\Omega \, d\omega^{\mathrm{intc}}_{\partial\Omega}$$

for some density measure $\omega^{\mathrm{intc}}_{\partial\Omega}$ of the type as in Proposition 3.1. The missing condition $\Omega \Subset U$ is compensated by a compact support of the χ_k^{intc} of the approximating sequence. However, a uniform bound on $\mathcal{H}^{n-1}(\partial\Omega_{-\delta})$ for small $\delta > 0$ similar to (4.82), that excludes certain sets of finite perimeter, is needed. But notice that the bound implies that the extension $\hat{F}$ of any $F \in \mathcal{DM}^\infty(\Omega)$ by zero has divergence measure on $\mathbb{R}^n$ (cf. [12, pp. 237, 242], [31, p. 150]). Thus the problem is equivalent to that for $\hat{F}$ with $\Omega \Subset \hat{U}$ for some open $\hat{U}$.

Proof of Proposition 4.34 For (1) we start with the case of the interior normal measure $\tilde{\nu}^{\mathrm{int}}$ and use

$$\chi := \tilde{\chi}^{\mathrm{int}} = \chi_{\mathrm{int}_*\Omega} \quad \text{and} \quad \chi_k := \tilde{\chi}_k^{\mathrm{int}}$$

(cf. Example 4.21). The first equation in (4.66) follows from Corollary 4.29 with

$$\int_{\partial\Omega} \chi \, d \operatorname{div}(\varphi F) + \operatorname{div}(\varphi F)(\operatorname{int}\Omega) = \operatorname{div}(\varphi F)(\operatorname{int}_*\Omega) .$$

Obviously $F\tilde{\nu}^{\mathrm{int}} \in \mathrm{ba}\big(U, \mathcal{B}(U), \mathcal{L}^n\big)^n$ and $\varphi \in \mathcal{W}^{1,\infty}(U)$ is continuous on $\overline{\Omega}$. Then, by Proposition 2.13, there is some related Radon measure $\tilde{\sigma}_F^{\mathrm{int}}$. Since $\chi_k \in \mathcal{W}^{1,\infty}(U)$ has compact support, the definition of divergence measure for $\chi_k F$ gives

$$\int_U \varphi F \cdot D\chi_k \, d\mathcal{L}^n = - \int_U \chi_k \varphi \, d \operatorname{div} F - \int_U \chi_k F \cdot D\varphi \, d\mathcal{L}^n$$

for all $\varphi \in C^1(U)$. By $\chi_k \to \chi_{\mathrm{int}_*\Omega}$ $\mathcal{H}^{n-1}$-a.e. and $\operatorname{div} F <<^{w} \mathcal{H}^{n-1}$ (cf. [41, p. 21]),

$$\lim_{k\to\infty} \int_U \varphi F \cdot D\chi_k \, d\mathcal{L}^n = - \int_{\mathrm{int}_*\Omega} \varphi \, d \operatorname{div} F - \int_{\mathrm{int}_*\Omega} F \cdot D\varphi \, d\mathcal{L}^n .$$

Hence, we do not need a subsequence in (4.33) and get

$$\lim_{k\to\infty} \int_U \varphi F \cdot D\chi_k \, d\mathcal{L}^n = - \fint_{\partial\Omega} \varphi F \, d\tilde{\nu}^{\mathrm{int}} = - \int_{\partial\Omega} \varphi \, d\tilde{\sigma}_F^{\mathrm{int}} .$$

Since $C_c^1(U)$ is dense in $C_c(U)$ we have $-F \cdot D\chi_k \mathcal{L}^n \overset{*}{\rightharpoonup} \tilde{\sigma}_F^{\mathrm{int}}$ as weak* limit of Radon measures. With the first equation in (4.66) and the results in [13, pp. 194, 200] we get the second equation in (4.75). For ψ_k from (4.42) we also have $2\chi_{\Omega^c} D\psi_k \mathcal{L}^n \overset{*}{\rightharpoonup} D\chi_\Omega$ by [13, p. 189] and $|D\psi_k|\mathcal{L}^n \overset{*}{\rightharpoonup} |D\chi_\Omega|$ by (4.44). Recall also (4.43). Then, for given $\varepsilon > 0$ and any open $B \Subset U$ with $|D\chi_\Omega|(\partial B) = 0$, there is some $\varphi_\varepsilon \in C_c(B)$ with $\|\varphi_\varepsilon\|_\infty \le 1$ such that, with arguments similar as in Comi-Payne [13, p. 196],

$$\begin{aligned}
&|\tilde{\sigma}_F^{\mathrm{int}}|(B) - \varepsilon \\
&\le \int_{\partial\Omega\cap B} \varphi_\varepsilon \, d\tilde{\sigma}_F^{\mathrm{int}} = - \lim_{k\to\infty} \int_{U\cap B} \varphi_\varepsilon F \cdot D\chi_k \, d\mathcal{L}^n \\
&= - \lim_{k\to\infty} 2 \int_{U\cap B\cap\Omega} \varphi_\varepsilon F \cdot D\psi_k \, d\mathcal{L}^n
\end{aligned}$$

$$
\begin{aligned}
&\le 2\|F\|_{\mathcal{L}^\infty(B)} \lim_{k\to\infty} \int_{U\cap B\cap\Omega} |D\psi_k|\, d\mathcal{L}^n \\
&= 2\|F\|_{\mathcal{L}^\infty(B)} \lim_{k\to\infty} \Big(\int_{U\cap B} |D\psi_k|\, d\mathcal{L}^n - \int_{U\cap B\cap\Omega^c} |D\psi_k|\, d\mathcal{L}^n \Big) \\
&\le 2\|F\|_{\mathcal{L}^\infty(B)} \lim_{k\to\infty} \Big(\int_{U\cap B} |D\psi_k|\, d\mathcal{L}^n - \Big| \int_{U\cap B} \chi_{\Omega^c} D\psi_k\, d\mathcal{L}^n \Big| \Big) \\
&= 2\|F\|_{\mathcal{L}^\infty(B)} \big(|D\chi_\Omega|(B) - \tfrac{1}{2} |D\chi_\Omega(B)| \big) \\
&= 2\|F\|_{\mathcal{L}^\infty(B)} (\mathcal{H}^{n-1} \llcorner \partial_*\Omega)(B) \big(1 - \tfrac{1}{2} \tfrac{|D\chi_\Omega(B)|}{|D\chi_\Omega|(B)} \big)
\end{aligned}
\tag{4.84}
$$

(for $|D\chi_\Omega|(B) \neq 0$ in the last line). By the arbitrariness of $\varepsilon > 0$ we can remove it. Then

$$
|\tilde{\sigma}_F^{\text{int}}|(B) \le 2\|F\|_{\mathcal{L}^\infty(B)} (\mathcal{H}^{n-1} \llcorner \partial_*\Omega)(B) .
$$

Since $\mathcal{H}^{n-1} \llcorner \partial_*\Omega$ is a Radon measure, we conclude that

$$
\operatorname{supp} \tilde{\sigma}_F^{\text{int}} \subset \partial_*\Omega \quad \text{and} \quad \tilde{\sigma}_F^{\text{int}} <<^w \mathcal{H}^{n-1} \llcorner \partial_*\Omega .
$$

Thus, the Radon-Nikodym theorem implies that $\tilde{\sigma}_F^{\text{int}}$ has an integrable density f^{int}. For fixed $x \in \partial^*\Omega$ and balls $B_r(x)$ we have that $|D\chi_\Omega|(\partial B_r(x)) = 0$ for $\mathcal{L}^1$-a.e. $r > 0$ and thus, (4.84) is valid for such $B = B_r(x)$. Then, by the definition of the reduced boundary, the fraction in (4.84) tends to 1 as $r \downarrow 0$ (cf. [2, p. 154]). Hence, by Lebesgue's differentiation theorem,

$$
f^{\text{int}} \in \mathcal{L}^\infty(\partial_*\Omega, \mathcal{H}^{n-1} \llcorner \partial_*\Omega) \quad \text{with} \quad \|f^{\text{int}}\|_\infty \le \|F\|_{\partial\Omega}
\tag{4.85}
$$

and, consequently,

$$
\int_{\partial\Omega} \varphi\, d\tilde{\sigma}_F^{\text{int}} = \int_{\partial_*\Omega} \varphi\, d\tilde{\sigma}_F^{\text{int}} = \int_{\partial_*\Omega} \varphi f^{\text{int}}\, d\mathcal{H}^{n-1}
$$

for all $\varphi \in \mathcal{W}^{1,\infty}(U)$. This verifies the first case. We can argue analogously in the second case by using that $\chi = \chi_{\partial_*\Omega \cup \text{int}_*\Omega}$ implies

$$
\int_{\partial\Omega} \chi\, d\, \text{div}(\varphi F) + \text{div}(\varphi F)(\text{int}\,\Omega) = \text{div}(\varphi F)(\partial_*\Omega \cup \text{int}_*\Omega)
$$

and that $\chi_k := \tilde{\chi}_k^{\text{ext}} \to \chi_{\partial_*\Omega \cup \text{int}_*\Omega}$ $\mathcal{H}^{n-1}$-a.e. on U. If Ω is open we argue as in Remark 4.36 to get (4.68). If Ω is closed we start with (4.67), we use

$$
\Omega = \text{int}_*\Omega \cup \partial_*\Omega \cup (\text{ext}_*\Omega \cap \partial\Omega), \quad \mathcal{L}^n(\text{ext}_*\, \Omega \cap \Omega) = 0 ,
$$

and we argue analogously to get (4.69) (cf. also [22, p. 222]).

For (2) we first observe that one always has the disjoint decomposition

$$\mathbb{R}^n = \mathrm{int}_*\Omega \cup \partial_*\Omega \cup \mathrm{ext}_*\Omega\,.$$

With $\partial_*\Omega \subset \partial\Omega$ we get

$$\partial_*\Omega \cup (\mathrm{int}_*\Omega \cap \partial\Omega) = \partial\Omega \backslash \mathrm{ext}_*\Omega \tag{4.86}$$

(cf. [32, pp. 49, 50]). Since $\mathcal{L}^n(\Omega) > 0$ and $\mathcal{H}^{n-1}(\partial\Omega \cap \mathrm{int}_*\Omega) < \infty$ we can apply Chen-Li-Torres [12, Theorem 3.1] to get sets $\Omega_k \Subset \Omega$ with finite perimeter such that

$$\sup_{k\in\mathbb{N}} \mathcal{H}^{n-1}(\partial_*\Omega_k) < \infty \quad \text{and} \quad \partial\Omega_k \subset (\partial\Omega)_{\frac{1}{2k}} \quad \text{for all } k \in \mathbb{N}$$

where the set inclusion is not explicitly stated there but it follows from the proof (cf. also [13, p. 208]). Then we define

$$\chi_k := \chi_{\Omega_k} * \eta_{\delta_k} \quad \text{with some} \quad \delta_k \in \big(0, \tfrac{1}{2k}\big)$$

where η_ε is the standard mollifier supported on $B_\varepsilon(0)$ (cf. also (4.42)). Obviously we have $\mathrm{supp}\,\chi_k \subset (\partial\Omega)_{\frac{1}{k}}$ and, by $\Omega_k \Subset \Omega$, we can choose $\delta_k > 0$ so small that even $\mathrm{supp}\,\chi_k \Subset \Omega$. Moreover $\chi_k \in W^{1,\infty}(U)$ with

$$\chi_k \to 1 \text{ on } \Omega\,, \qquad \chi_k \to 0 \text{ otherwise.}$$

Since $|D\chi_{\Omega_k}|(\partial U) = 0$ and by (4.46) with Ω_k instead of Ω, we can apply (4.46) to Ω_k. Therefore we can assume $\delta_k > 0$ to be so small that

$$|D\chi_k|(U) = \int_U |D\chi_k|\, d\mathcal{L}^n \le |D\chi_{\Omega_k}|(U) + 1 = \mathcal{H}^{n-1}(\partial_*\Omega_k) + 1\,.$$

Consequently, we have that $\{\chi_k\}$ is an approximating sequence for the good approximation $\chi = \chi_{\mathrm{int}\,\Omega}$ of χ_Ω and each $(\partial\Omega)_{\frac{1}{k}} \cap \Omega$ is an aura of the associated normal measure ν (cf. Theorem 4.19). Then (4.70) follows from Theorem 4.27 and Corollary 4.29.

By $\mathcal{H}^{n-1}(\partial\Omega \cap \mathrm{int}_*\Omega) < \infty$ we have $\hat{F} \in \mathcal{DM}^\infty(\mathbb{R}^n)$ for the extension $\hat{F}$ of F by zero (cf. [12, p. 242]). Let us choose some open bounded $\hat{U} \subset \mathbb{R}^n$ such that $\Omega \Subset \hat{U}$. Then we can apply assertion (1) to get (4.66) for $\hat{F}$ and all $\varphi \in C^1(\mathbb{R}^n)$. According to Remark 4.36 this can be transformed to (4.79) and, by $\mathrm{div}\, F = \mathrm{div}(\hat{F})$ on Ω, we get

$$\mathrm{div}(\varphi F)(\Omega) = \int_{\partial\Omega\backslash \mathrm{ext}_*\Omega} \varphi\, d\sigma_F$$

for some Radon measure σ_F supported on $\partial\Omega \setminus \mathrm{ext}_*\Omega$. The assumptions and (4.86) imply that $\mathcal{H}^{n-1}(\partial\Omega \setminus \mathrm{ext}_*\Omega) < \infty$. Therefore σ_F has an $\mathcal{H}^{n-1}$-integrable density f by the Radon-Nikodym theorem and we obtain (4.71) (cf. the arguments following (4.79)). Then (4.72) follows from (4.78). In order to show that f is essentially bounded we argue as in Chen-Li-Torres [12, p. 250] but without smooth approximation of F. We fix some $B_r(x)$ with $x \in \partial\Omega \setminus \mathrm{ext}_*\Omega$ and $r > 0$. Moreover we observe that the sets $\Omega_k \Subset \Omega$ from above can be chosen such that they have smooth boundary, and can therefore assumed to be open, and that

$$|D\chi_{\Omega_k}|(B_r(x)) = (\mathcal{H}^{n-1} \llcorner \partial_*\Omega_k)(B_r(x)) \leq c\mathcal{H}^{n-1}\big((\partial\Omega \setminus \mathrm{ext}_*\Omega) \cap B_r(x)\big)$$

for some $c > 0$ depending merely on dimension n (cf. [12, p. 237, (5.8)] and also [13, p. 208]). Now, for some $\varphi \in C^1(\mathbb{R}^n)$ with $\|\varphi\|_\infty \leq 1$ and supported on $B_r(x)$, we use dominated convergence, $\Omega_k = \mathrm{int}_*\Omega_k$, and assertion (1) for Ω_k with density functions f_k to get

$$\begin{aligned}
\int_{\partial\Omega\setminus\mathrm{ext}_*\Omega} \varphi f \, d\mathcal{H}^{n-1} &= \mathrm{div}(\varphi F)(\Omega) \\
&= \int_\Omega \varphi \, d \, \mathrm{div}\, F + \int_\Omega F \cdot D\varphi \, d\mathcal{L}^n \\
&= \lim_{k\to\infty} \Big(\int_{\Omega_k} \varphi \, d \, \mathrm{div}\, F + \int_{\Omega_k} F \cdot D\varphi \, d\mathcal{L}^n \Big) \\
&= \lim_{k\to\infty} \int_{\partial_*\Omega_k} \varphi f_k \, d\mathcal{H}^{n-1} \\
&\leq \lim_{k\to\infty} \|F\|_{\mathcal{L}^\infty((\partial\Omega)_{1/k})} \int_{\partial_*\Omega_k} |\varphi| \, d\mathcal{H}^{n-1} \\
&\leq \|F\|_{\partial\Omega} \lim_{k\to\infty} (\mathcal{H}^{n-1} \llcorner \partial_*\Omega_k)(B_r(x)) \\
&\leq c\|F\|_{\partial\Omega} \mathcal{H}^{n-1}\big((\partial\Omega \setminus \mathrm{ext}_*\Omega) \cap B_r(x)\big) .
\end{aligned}$$

Hence, by Lebesgue's differentiation theorem, $\|f\|_{\partial\Omega} \leq c\|F\|_{\partial\Omega}$.

If $\hat{F}$ is continuous, we use $\chi_{\mathrm{int}\,\Omega} = \chi_\Omega$ to get (4.34) with the normal measure ν derived above (that we assume to be extended with zero on $\mathbb{R}^n$). Then (4.70) and (4.34) with the vector function φF give the second equality in (4.73). Since we also have (4.71), the measure $f\mathcal{H}^{n-1} \llcorner (\partial\Omega \setminus \mathrm{ext}_*\Omega)$ has to vanish outside $\partial_*\Omega$. By (4.86) this means that $(\mathrm{div}\, F)(\partial\Omega \cap \mathrm{int}_*\Omega) = 0$ (cf. also [13, p. 198]). Using (4.77) we get the first equality in (4.73). □

Next we consider an example with high oscillations that is occasionally discussed in the literature (cf. Chen-Frid [6, p. 403], Chen-Torres-Ziemer [10, p. 258], Comi-Payne [13, p. 216]).

Example 4.39 Let $\Omega = \{y < x\} \cap \{|x| + |y| < 1\} \subset \mathbb{R}^2$ and $F : \mathbb{R}^2 \to \mathbb{R}^2$ given by

$$F(x, y) = \left(\sin\left(\tfrac{1}{x-y}\right), \sin\left(\tfrac{1}{x-y}\right)\right) \quad \text{for } x \neq y\,.$$

One has $\operatorname{div} F = 0$ on $\mathbb{R}^2$ in the distributional sense and thus $F \in \mathcal{DM}^\infty(\Omega)$. But the component functions of F do not belong to $\mathcal{BV}(\Omega)$ and they do not have a classical trace on the line $\{x = y\}$. From Proposition 4.30 we get

$$\operatorname{div}(\varphi F)(\Omega) = ⨍_{\partial\Omega} \varphi F \cdot \nu^\Omega \, d\omega_{\partial\Omega}^{\text{int}} \tag{4.87}$$

for all $\varphi \in \mathcal{W}^{1,\infty}(\Omega)$. The measure $\omega_{\partial\Omega}^{\text{int}}$ satisfies (4.63) and is some pure extension of $\mathcal{H}^1 \lfloor \partial\Omega$ on Ω according to Proposition 2.13 (cf. also below). A trace function of F on $\partial\Omega$ is not needed. We can divide Ω by its diagonals into two large triangles Ω_j^l and two small triangles Ω_j^s $(j = 1, 2)$. Then we have

$$F \cdot \nu^\Omega = 0 \text{ on the } \Omega_j^l \quad \text{and} \quad F \cdot \nu^\Omega = \pm 2 \sin\left(\tfrac{1}{x-y}\right) \text{ on the } \Omega_j^s.$$

Therefore we can disregard the large sides in (4.87) and get

$$\operatorname{div}(\varphi F)(\Omega) = ⨍_{\partial\Omega\cap\partial\Omega_1^s} \varphi F \cdot \nu^\Omega \, d\omega_{\partial\Omega}^{\text{int}} + ⨍_{\partial\Omega\cap\partial\Omega_2^s} \varphi F \cdot \nu^\Omega \, d\omega_{\partial\Omega}^{\text{int}}$$

(notice that the boundaries of the triangles Ω_j^s, Ω_j^l within Ω can be neglected, since $\omega_{\partial\Omega}^{\text{int}}$ is weakly absolutely continuous with respect to $\mathcal{L}^2$). By $\operatorname{div} F = 0$ the left hand side vanishes for constant φ. Moreover, using (4.63), the σ-measure σ_F associated to $F \cdot \nu^\Omega \omega_{\partial\Omega}^{\text{int}}$ according to Proposition 2.13 is supported on $\partial\Omega$ and we have

$$\begin{aligned} &\sigma_F = 0 \text{ on } \partial\Omega \cap \partial\Omega_j^l \quad \text{and} \\ &\sigma_F = \pm 2 \sin\left(\tfrac{1}{x-y}\right)\mathcal{H}^1 \text{ on } \partial\Omega \cap \partial\Omega_j^s \quad (j = 1, 2)\,. \end{aligned}$$

Hence

$$\operatorname{div}(\varphi F)(\Omega) = 2 \int_{\partial\Omega\cap\partial\Omega_1^s} \varphi \sin\left(\tfrac{1}{x-y}\right) d\mathcal{H}^1 - 2 \int_{\partial\Omega\cap\partial\Omega_2^s} \varphi \sin\left(\tfrac{1}{x-y}\right) d\mathcal{H}^1 \tag{4.88}$$

for all $\varphi \in C^1(\overline{\Omega})$. This means that the highly oscillating F has a normal trace function on $\partial\Omega$ and we get a Gauss-Green formula as already derived in [10] and [13] (cf. also Proposition 4.34 for a direct statement). We can argue the same way for $\overline{\Omega}$ with $\omega_{\partial\Omega}^{\text{ext}}$. Notice that the problem can be treated analogously for other Ω touching the set $\{x = y\}$.

4.3 Sobolev Functions and BV Functions

In this section we show that the previous results are applicable to Sobolev functions and BV functions. Recall that, for $f \in \mathcal{BV}(U)$ and $\Omega \in \mathcal{B}(U)$,

$$\varphi \to \operatorname{div}(\varphi f)(\Omega) = \int_{\Omega} f \operatorname{div} \varphi \, d\mathcal{L}^n + \int_{\Omega} \varphi \, dDf$$

is a trace on $\partial\Omega$ over $\mathcal{W}^{1,\infty}(U, \mathbb{R}^n)$ according to Proposition 3.7. As direct consequence of Theorem 4.1 we provide a general Gauss-Green formula for BV functions by arguing as in the proof of Proposition 3.7. Notice that we have to take $m = n$ for the particular cases (G), (L) and (C) in this section (recall the related definitions of X_0 after Proposition 3.13).

Theorem 4.40 *Let $U \subset \mathbb{R}^n$ be open and bounded, let $\Omega \in \mathcal{B}(U)$, let $\delta > 0$, and let $f \in \mathcal{BV}(U)$. Then there exist measures*

$$\lambda_f \in \operatorname{ba}\big(U, \mathcal{B}(U), \mathcal{L}^n\big)^n \quad \textit{and} \quad \mu_f \in \operatorname{ba}\big(U, \mathcal{B}(U), \mathcal{L}^n\big)^{n \times n}$$

with $\operatorname{core} \lambda_f$, $\operatorname{core} \mu_f \subset \overline{(\partial\Omega)_\delta \cap U}$ *such that*

$$\langle Tf, \varphi \rangle = \operatorname{div}(\varphi f)(\Omega) = \int_{(\partial\Omega)_\delta \cap U} \varphi \, d\lambda_f + \int_{(\partial\Omega)_\delta \cap U} D\varphi \, d\mu_f \tag{4.89}$$

for all $\varphi \in \mathcal{W}^{1,\infty}(U, \mathbb{R}^n)$ with $T : \mathcal{BV}(U) \to \mathcal{W}^{1,\infty}(U, \mathbb{R}^n)^*$ from Proposition 3.7. In the particular cases with $\Gamma = \partial\Omega$ we have in addition that:

(L): *λ_f, μ_f can be chosen such that* $\operatorname{core} \lambda_f \subset \partial\Omega$ *and, thus,* (4.89) *becomes*

$$\operatorname{div}(\varphi f)(\Omega) = ⨍_{\partial\Omega} \varphi \, d\lambda_f + \int_{(\partial\Omega)_\delta \cap U} D\varphi \, d\mu_f \, .$$

(C): *λ_f, μ_f can be chosen such that λ_f corresponds to a Radon measure σ_f with* $\operatorname{supp} \sigma_f \subset \partial\Omega$ *and, thus,*

$$\operatorname{div}(\varphi f)(\Omega) = \int_{\partial\Omega} \varphi \, d\sigma_f + \int_{(\partial\Omega)_\delta \cap U} D\varphi \, d\mu_f$$

(where we always identify φ with its continuous extension on $\partial\Omega$).

We call (λ_f, μ_f), representing an element of $\mathcal{W}^{1,\infty}(U, \mathbb{R}^n)^*$, *normal trace* of f near $\partial\Omega$. Notice that $D\varphi \, d\mu_f$ in (4.89) has to be taken as scalar product of matrices. Since $\mathcal{W}^{1,1}(U) \subset \mathcal{BV}(U)$, the result covers these Sobolev functions (cf. also Remark 3.8).

Proof For vector fields $F_k \in \mathcal{DM}^1(U)$ related to $f \in \mathcal{BV}(U)$ as in the proof of Proposition 3.7, we can apply Theorem 4.1. This way we get measures

$$\lambda_{F_k} \in \mathrm{ba}\big(U, \mathcal{B}(U), \mathcal{L}^n\big) \quad \text{and} \quad \mu_{F_k} \in \mathrm{ba}\big(U, \mathcal{B}(U), \mathcal{L}^n\big)^n$$

with $\operatorname{core}\lambda_{F_k}$, $\operatorname{core}\mu_{F_k} \subset \overline{(\partial\Omega)_\delta \cap U}$ and such that we have for every vector field $\varphi = (\varphi^1, \dots, \varphi^n) \in \mathcal{W}^{1,\infty}(U, \mathbb{R}^n)$

$$\big(D_{x_k}(\varphi^k f)\big)(\Omega) = \int_{(\partial\Omega)_\delta \cap U} \varphi^k \, d\lambda_{F_k} + \int_{(\partial\Omega)_\delta \cap U} D\varphi^k \, d\mu_{F_k}$$

with the scalar measure

$$D_{x_k}(\varphi^k f) = f D_{x_k}\varphi^k \mathcal{L}^n + \varphi^k D_{x_k} f$$

(cf. Remark 3.8 and [2, p. 118]). The sum over k gives the first statement and the particular cases follow directly from these in Theorem 4.1. □

As in Proposition 4.5 we can characterize the cases where $\mu_f = 0$ is possible and where the measures λ_f, μ_f can be chosen independent of δ.

Proposition 4.41 *Let $U \subset \mathbb{R}^n$ be open and bounded, let $\Omega \in \mathcal{B}(U)$, and assume that $f \in \mathcal{BV}(U)$.*

(1) *In Theorem* 4.40 *we can choose (λ_f, μ_f) with $\operatorname{core}\lambda_f$, $\operatorname{core}\mu_f \subset \partial\Omega$, i.e. λ_f, μ_f are independent of δ, if and only if*

$$\liminf_{\delta \downarrow 0} \sup_{\substack{\varphi \in \mathcal{W}^{1,\infty}(U,\mathbb{R}^n) \\ \|\varphi_{|(\partial\Omega)_\delta \cap U}\|_{\mathcal{W}^{1,\infty}} \le 1}} \operatorname{div}(\chi_\delta^{\partial\Omega} \varphi f)(\Omega) < \infty \tag{4.90}$$

with $\chi_\delta^{\partial\Omega}$ as in (3.31). *In this case* (4.89) *becomes*

$$\operatorname{div}(\varphi f)(\Omega) = \fint_{\partial\Omega} \varphi \, d\lambda_f + \fint_{\partial\Omega} D\varphi \, d\mu_f \,. \tag{4.91}$$

(2) *In Theorem* 4.40 *we can choose (λ_f, μ_f) for $\delta > 0$ such that $\mu_f = 0$ and $\operatorname{core}\lambda_f \subset \overline{(\partial\Omega)_\delta \cap U}$ if and only if*

$$\sup_{\substack{\varphi \in \mathcal{W}^{1,\infty}(U,\mathbb{R}^n) \\ \|\varphi_{|(\partial\Omega)_\delta \cap U}\|_{\mathcal{L}^\infty} \le 1}} \operatorname{div}(\varphi f)(\Omega) < \infty \,. \tag{4.92}$$

(3) *In Theorem* 4.40 *we can take* (λ_f, μ_f) *with* $\mu_f = 0$ *and* $\operatorname{core}\lambda_f \subset \partial\Omega$ *if and only if*

$$\liminf_{\delta\downarrow 0} \sup_{\substack{\varphi\in\mathcal{W}^{1,\infty}(U,\mathbb{R}^n)\\ \|\varphi_{|(\partial\Omega)_\delta\cap U}\|_{\mathcal{L}^\infty}\le 1}} \operatorname{div}(\varphi f)(\Omega) < \infty\,.$$

Proof We use the notation of the proof of Theorem 4.40. We have for all k that

$$\chi_\delta^{\partial\Omega}\varphi^k F_k = \chi_\delta^{\partial\Omega}\varphi f \quad \text{if} \quad \varphi^j = 0 \quad \text{for all } j \neq k\,.$$

Hence (4.90) is equivalent to (4.5) for all F_k and, by Proposition 4.5, this is equivalent to $\operatorname{core}\lambda_{F_k}$, $\operatorname{core}\mu_{F_k} \subset \partial\Omega$ for all k. But this gives (1). Analogously we derive (2) and (3) from Proposition 4.5. □

Arguing as in the previous proof we can transfer Lemma 4.6, Proposition 4.7, and Proposition 4.8 to BV functions. Let us briefly rephrase these results for completeness and for the convenience of the reader.

Lemma 4.42 *Condition* (4.90) *in Proposition* 4.41 *is equivalent to each of the following two conditions*

$$\liminf_{\delta\downarrow 0} \sup_{\substack{\varphi\in\mathcal{W}^{1,\infty}(U,\mathbb{R}^n)\\ \|\varphi_{|(\partial\Omega)_\delta\cap U}\|_{\mathcal{W}^{1,\infty}}\le 1}} \int_{(\partial\Omega)_\delta\cap\Omega} \varphi f D\chi_\delta^{\partial\Omega}\, d\mathcal{L}^n < \infty\,,$$

$$\liminf_{\delta\downarrow 0} \sup_{\substack{\varphi\in\mathcal{W}^{1,\infty}(U,\mathbb{R}^n)\\ \|\varphi_{|(\partial\Omega)_\delta\cap U}\|_{\mathcal{W}^{1,\infty}}\le 1}} \frac{1}{\delta}\int_{((\partial\Omega)_\delta\setminus(\partial\Omega)_{\delta/2})\cap\Omega} \varphi f D\operatorname{dist}_{\partial\Omega}\, d\mathcal{L}^n < \infty\,.$$

Moreover, $\operatorname{int}\Omega = \emptyset$ *implies* (4.90).

Proposition 4.43 *Let* $U \subset \mathbb{R}^n$ *be an open and bounded set, let* $\Omega \in \mathcal{B}(U)$ *satisfy* $\mathcal{L}^n(\Omega\setminus\operatorname{int}\Omega) = 0$, *and assume that* $f \in \mathcal{BV}(U)$. *If*

$$\liminf_{\delta\downarrow 0} \int_\Omega \left|f D\chi_\delta^{\partial\Omega}\right| d\mathcal{L}^n < \infty\,, \tag{4.93}$$

then we can chose $\mu_f = 0$ *and* λ_f *with* $\operatorname{core}\lambda_f \subset \partial\Omega$ *in Theorem* 4.40. *We have* (4.93) *if* f *is bounded and if there is some* $\tilde\delta > 0$ *such that*

$$\sup_{\delta\in(0,\tilde\delta)} \mathcal{H}^{n-1}(\partial\Omega_{-\delta}) < \infty\,. \tag{4.94}$$

If Ω *has finite perimeter, then* (4.12) *for some* $c, r > 0$ *ensures* (4.94).

Proposition 4.44 *Let $U \subset \mathbb{R}^n$ be open and bounded and let $f \in \mathcal{BV}(U)$.*

(1) *If $\Omega \in \mathcal{B}(U)$ is such that any $\varphi \in \mathcal{W}^{1,\infty}(U)$ has a continuous extension onto $\overline{\Omega}$, if $\mathcal{L}^n(\Omega \setminus \operatorname{int}\Omega) = 0$, and if there are $c > 0$ and $\tilde{\delta} > 0$ such that*

$$\|\varphi_{|\partial\Omega}\|_{\mathrm{Lip}(\partial\Omega)} \le c\|\varphi\|_{\mathcal{W}^{1,\infty}((\partial\Omega)_\delta \cap U)} \quad \text{for all} \quad \varphi \in \mathcal{W}^{1,\infty}(U),\ \delta \in (0,\tilde{\delta}), \tag{4.95}$$

then (4.90) *is satisfied.*

(2) *If Ω is open with Lipschitz boundary, then* (4.90) *is satisfied.*

Notice that we do not have to consider $\varphi \in \mathcal{W}^{1,\infty}(U, \mathbb{R}^n)$ in the previous proposition.

Simple examples of Gauss-Green formulas for Sobolev functions going beyond the classical ones can be obtained from Example 4.10 or 4.11 if we take the first component of the vector field F as function $f \in \mathcal{W}^{1,1}(U)$. Let us now provide a Sobolev function on a set $\Omega \subset \mathbb{R}^2$ of finite perimeter where the precise representative is not $\mathcal{H}^1$-integrable on $\partial\Omega$. This certainly prevents a usual Gauss-Green formula. By the derivation of measures λ_f and μ_f for (4.89) we demonstrate how more general Gauss-Green formulas can be obtained.

Example 4.45 Using the unit square

$$\tilde{\Omega} := \{(x,y) \in \mathbb{R}^2 \mid x, y \in (0,1)\},$$

$$x_k := \tfrac{1}{k},\ y_k := \left(\tfrac{1}{k}\right)^{\frac{5}{4}}, \quad \text{segments} \quad \Gamma_k := \{x_k\} \times [0, y_k],$$

and the closed convex sets

$$\tilde{\Omega}_k := \operatorname{conv}\{\Gamma_{2k}, \Gamma_{2k+1}\},$$

we define the open sets

$$U := \Omega := \tilde{\Omega} \setminus \bigcup_{k=1}^{\infty} \tilde{\Omega}_k. \tag{4.96}$$

Since $\sum_k y_k$ is finite, Ω has finite perimeter. Let us consider $f \in \mathcal{W}^{1,1}(\Omega)$ with

$$f(x,y) := \frac{1}{|(x,y)|_\infty^{1/4}}$$

where $|\cdot|_\infty$ is the ∞-norm (one even has $f \in \mathcal{W}^{1,p}(\Omega)$ for $1 \le p < \frac{8}{5}$). Clearly

$$\int_{\partial\Omega} |f|\, d\mathcal{H}^1 \ge \int_{\bigcup_k \Gamma_k} |f|\, d\mathcal{H}^1 = \sum_k \left(\tfrac{1}{k}\right)^{\frac{5}{4}} k^{\frac{1}{4}} = \sum_k \tfrac{1}{k}$$

and thus, $f \notin \mathcal{L}^1(\partial\Omega, \mathcal{H}^1)$. Therefore we do not have a Gauss-Green formula in the classical sense.

Nevertheless we can apply Theorem 4.40 and (4.89) is valid with measures λ_f and μ_f. By local arguments near the points $x \in \partial\Omega$ (in particular near the origin), we obtain that any $\varphi \in \mathcal{W}^{1,\infty}(\Omega, \mathbb{R}^2)$ can be extended continuously to $\partial\Omega$. Since $(\partial\Omega)_\delta \cap \Omega$ is bounded path connected with $\partial\Omega$ for any $\delta > 0$, we have case (C) by Proposition 3.17. Thus λ_f can be chosen as Radon measure independent of δ. We now show by construction, without verifying (4.90), that there is a vector-valued Radon measure σ_f supported on $\partial\Omega$ and a measure $\mu_f \in \mathrm{ba}(\Omega, \mathcal{B}(\Omega), \mathcal{L}^2)^{2\times 2}$ with core $\mu_f \subset \partial\Omega$ such that

$$\mathrm{div}(\varphi f)(\Omega) = \int_{\partial\Omega} \varphi \, d\sigma_f + \int\!\!\!\!\!\!-_{\partial\Omega} D\varphi \, d\mu_f$$

for all $\varphi \in \mathcal{W}^{1,\infty}(\Omega, \mathbb{R}^2)$ (identified with their extension onto $\overline{\Omega}$).

Let us analyze how σ_f and μ_f, that are not unique, could look like. First we restrict our attention to smooth $\varphi = (\varphi_1, \varphi_2) \in C^1(\mathbb{R}^2, \mathbb{R}^2)$ such that we can also consider μ_f as Radon measure (cf. Proposition 2.13). Notice that $\tilde{\Omega}_k$ has Lipschitz boundary and we decompose

$$\partial\tilde{\Omega}_k = \Gamma_{2k} \cup \Gamma_{2k+1} \cup \Gamma_k^0 \cup \Gamma_k^1$$

where Γ_k^0 is the part on the x-axis and Γ_k^1 is the opposite part. Then, taking $\mathrm{div}(\varphi f)$ as a measure, we get

$$\mathrm{div}(\varphi f)(\Omega) = \mathrm{div}(\varphi f)(\tilde{\Omega}) - \sum_{k=1}^{\infty} \mathrm{div}(\varphi f)(\tilde{\Omega}_k) \tag{4.97}$$

for all $\varphi \in C^1(\mathbb{R}^2, \mathbb{R}^2)$. Obviously $f \in \mathcal{L}^1(\partial\tilde{\Omega}, \mathcal{H}^1)$ and, by the usual Gauss-Green formula,

$$\mathrm{div}(\varphi f)(\tilde{\Omega}) = \int_{\partial\tilde{\Omega}} f\varphi \cdot \nu^{\tilde{\Omega}} \, d\mathcal{H}^1 . \tag{4.98}$$

Moreover

$$\begin{aligned} \mathrm{div}(\varphi f)(\tilde{\Omega}_k) &= \int_{\partial\tilde{\Omega}_k} f\varphi \cdot \nu^{\tilde{\Omega}_k} \, d\mathcal{H}^1 \\ &= \int_{\Gamma_{2k}} f\varphi_1 \, d\mathcal{H}^1 - \int_{\Gamma_{2k+1}} f\varphi_1 \, d\mathcal{H}^1 + \int_{\Gamma_k^0 \cup \Gamma_k^1} f\varphi \cdot \nu^{\tilde{\Omega}_k} \, d\mathcal{H}^1 . \end{aligned} \tag{4.99}$$

If we plug (4.98) and (4.99) into (4.97), the integrals on Γ_k^0 cancel each other. Thus the integral in (4.98) has to be evaluated merely on the set $\tilde{\Gamma} := \partial\tilde{\Omega} \setminus \big(\bigcup_k \Gamma_k^0\big)$. Hence the Radon measure

$$\sigma_f^0 := f \nu^{\tilde{\Omega}} \mathcal{H}^1 \llcorner \tilde{\Gamma}$$

can be taken as part of σ_f. The integral on Γ_k^1 in (4.99) is related to the Radon measure

$$\sigma_f^{k1} := f \nu^{\tilde{\Omega}_k} \mathcal{H}^1 \llcorner \Gamma_k^1 .$$

Since the sum over $k \in \mathbb{N}$ is also a Radon measure, we can take it with the opposite sign also as part of σ_f. It remains to consider the integrals on Γ_k in (4.99). Here we have to realize that the sum of the Radon measures $f\mathcal{H}^1 \llcorner \Gamma_k$ is not bounded (cf. above) and thus, not a Radon measure. Therefore we have to transform these integrals on Γ_k in a suitable way (it would be sufficient to take finitely many of these measures with the correct sign as part of σ_f and to transform merely the rest, which would lead to measures μ_f that differ from those derived below). With

$$g(x, y) := x^{-\frac{1}{4}} y \,, \quad G^0(x, y) := \begin{pmatrix} 0 & -x^{-\frac{1}{4}} y \\ 0 & 0 \end{pmatrix},$$

integration by parts gives

$$\begin{aligned}
\int_{\Gamma_k} f\varphi_1 \, d\mathcal{H}^1 &= \int_0^{y_k} g_y(x_k, y)\varphi_1(x_k, y)\, dy \\
&= -\int_0^{y_k} g(x_k, y)\varphi_{1,y}(x_k, y)\, dy + \big[g(x_k, y)\varphi_1(x_k, y)\big]_0^{y_k} \\
&= -x_k^{-\frac{1}{4}} \int_0^{y_k} y\varphi_{1,y}(x_k, y)\, dy + g(x_k, y_k)\varphi_1(x_k, y_k) \\
&= \int_0^{y_k} G^0(x_k, y) : D\varphi(x_k, y)\, dy + \tfrac{1}{k}\varphi_1(x_k, y_k) \\
&= \int_{\Gamma_k} G^0 : D\varphi \, d\mathcal{H}^1 + \tfrac{1}{k}\varphi_1(x_k, y_k)
\end{aligned} \tag{4.100}$$

(here : denotes the scalar product of matrices). Now, taken with the correct sign, the measures $G^0\mathcal{H}^1 \llcorner \Gamma_k$ with total variation

$$\big|G^0\mathcal{H}^1 \llcorner \Gamma_k\big|(\Gamma_k) = \int_0^{y_k} x_k^{-\frac{1}{4}} y\, dy = \tfrac{1}{2}\big(\tfrac{1}{k}\big)^{\frac{9}{4}}$$

could contribute to μ_f and the Dirac measures $\frac{1}{k}\delta_{(x_k,y_k)}$ might contribute to σ_f. While the sum of the $G^0\mathcal{H}^1\lfloor\Gamma_k$ gives a finite measure, the sum of the Dirac measures is not finite and needs some further transformation. We therefore consider (cf. (4.99))

$$
\begin{aligned}
&\int_{\Gamma_{2k}} f\varphi_1\, d\mathcal{H}^1 - \int_{\Gamma_{2k+1}} f\varphi_1\, d\mathcal{H}^1 \\
&\qquad = \int_{\Gamma_{2k}} G^0 : D\varphi\, d\mathcal{H}^1 - \int_{\Gamma_{2k+1}} G^0 : D\varphi\, d\mathcal{H}^1 \\
&\qquad\qquad + \tfrac{1}{2k}\varphi_1(x_{2k}, y_{2k}) - \tfrac{1}{2k+1}\varphi_1(x_{2k+1}, y_{2k+1})\,.
\end{aligned}
$$

Using $\frac{1}{k} = \frac{1}{k+1} + \frac{1}{k(k+1)}$ and

$$
v_k := \frac{(\overset{\vartriangle}{x}_k, \overset{\vartriangle}{y}_k)}{\left|(\overset{\vartriangle}{x}_k, \overset{\vartriangle}{y}_k)\right|} \quad \text{with} \quad (\overset{\vartriangle}{x}_k, \overset{\vartriangle}{y}_k) := (x_{2k}, y_{2k}) - (x_{2k+1}, y_{2k+1})
$$

we get

$$
\begin{aligned}
&\tfrac{1}{2k}\varphi_1(x_{2k}, y_{2k}) - \tfrac{1}{2k+1}\varphi_1(x_{2k+1}, y_{2k+1}) \\
&\quad = \tfrac{1}{2k+1}\big(\varphi_1(x_{2k}, y_{2k}) - \varphi_1(x_{2k+1}, y_{2k+1})\big) + \tfrac{1}{2k(2k+1)}\varphi_1(x_{2k}, y_{2k}) \\
&\quad = \tfrac{1}{2k+1}\int_0^1 D\varphi_1(x_{2k+1} + t\,\overset{\vartriangle}{x}_k, y_{2k+1} + t\,\overset{\vartriangle}{y}_k)\cdot(\overset{\vartriangle}{x}_k, \overset{\vartriangle}{y}_k)\, dt \\
&\qquad + \tfrac{1}{2k(2k+1)}\varphi_1(x_{2k}, y_{2k}) \\
&\quad = \tfrac{1}{2k+1}\int_{\Gamma_k^1} v_k \cdot D\varphi_1\, d\mathcal{H}^1 + \tfrac{1}{2k(2k+1)}\varphi_1(x_{2k}, y_{2k})\,.
\end{aligned}
$$

Here the sum of the Dirac measures $\frac{1}{2k(2k+1)}\delta_{(x_{2k},y_{2k})}$ is finite and thus, it can contribute to σ_f. The measures $G_k^1\mathcal{H}^1\lfloor\Gamma_k^1$ with

$$
G_k^1(x, y) := \tfrac{1}{2k+1}\begin{pmatrix} v_{k,1} & v_{k,2} \\ 0 & 0 \end{pmatrix}
$$

have total variation

$$
\tfrac{1}{2k+1}\mathcal{H}^1(\Gamma_k^1) \le \tfrac{1}{2k+1}\left(\tfrac{3}{2k} - \tfrac{3}{2k+1}\right) = \tfrac{3}{2k(2k+1)^2}\,.
$$

Thus the sum of these measures is finite and can contribute to μ_f. Hence we get the desired Gauss-Green formula for $\varphi \in C^1(\mathbb{R}^2, \mathbb{R}^2)$ with Radon measures

$$\sigma_f := \sigma_f^0 - \sum_{k=1}^{\infty} \left(f \nu^{\tilde{\Omega}_k} \mathcal{H}^1 \lfloor \Gamma_k^1 + \tfrac{1}{2k(2k+1)} \delta_{(x_{2k}, y_{2k})} \right)$$

$$\mu_f := \sum_{k=1}^{\infty} G^0 \mathcal{H}^1 \lfloor \Gamma_{2k} - G^0 \mathcal{H}^1 \lfloor \Gamma_{2k+1} + G_k^1 \mathcal{H}^1 \lfloor \Gamma_k^1 \,.$$

Let us now consider the extension of μ_f to $\varphi \in \mathcal{W}^{1,\infty}(\Omega, \mathbb{R}^2)$. Since an extension by Hahn-Banach will not be unique in general, we have to argue more carefully. Instead of an approximation by smooth functions as in Example 4.14 we want to present a more direct argument. First we consider Γ_k with odd index k and set

$$\Gamma_{kt} := \left\{ (x_k + t, y) \;\middle|\; y \in [0, y_k] \right\}, \quad t \in \mathbb{R}\,.$$

Since f, φ are continuous on a neighborhood of Γ_k intersected with Ω and since $\varphi(x_k + t, \cdot)$ is absolutely continuous on Γ_{kt} for almost all small $t > 0$, we argue similar as for (4.100) to get

$$\begin{aligned} \int_{\Gamma_k} f \varphi_1 \, d\mathcal{H}^1 &= \lim_{t \downarrow 0} \frac{1}{t} \int_0^t \int_{\Gamma_{k\tau}} f \varphi_1 \, d\mathcal{H}^1 \, d\tau \\ &= \lim_{t \downarrow 0} \frac{1}{t} \int_0^t \left(\int_{\Gamma_{k\tau}} G^0 : D\varphi \, d\mathcal{H}^1 + \frac{y_k}{(x_k + \tau)^{1/4}} \varphi_1(x_k + \tau, y_k) \right) d\tau \\ &= \lim_{t \downarrow 0} \frac{1}{t} \int_0^t \int_{\Gamma_{k\tau}} G^0 : D\varphi \, d\mathcal{H}^1 \, d\tau + \tfrac{1}{k} \varphi_1(x_k, y_k)\,. \end{aligned}$$

Obviously

$$\Phi \to \lim_{t \downarrow 0} \frac{1}{t} \int_0^t \int_{\Gamma_{k\tau}} \Phi : G^0 \, d\mathcal{H}^1 \, d\tau$$

is a linear continuous functional on $\mathcal{L}^\infty(\Omega, \mathbb{R}^{2\times 2})$ depending on values of Φ near Γ_k. Hence there are matrix-valued measures $\mu_k^0 \in \mathrm{ba}(\Omega, \mathcal{B}(\Omega), \mathcal{L}^2)^{2\times 2}$ with core Γ_k such that

$$⨍_{\Gamma_k} \Phi \, d\mu_k^0 = \lim_{t \downarrow 0} \frac{1}{t} \int_0^t \int_{\Gamma_{k\tau}} G^0 : \Phi \, d\mathcal{H}^1 \, d\tau$$

for all $\Phi \in \mathcal{L}^\infty(\Omega, \mathbb{R}^{2\times 2})$. By the special form of G^0 the only non-vanishing component of the measure μ_k^0 is $(\mu_k^0)_{12}$. More precisely, with μ_{Γ_k} from Proposition 3.1 where $E = \Omega$ and $\gamma(\delta) = \delta$, we obtain from (3.5) that

$$(\mu_k^0)_{12} = \left(-x^{-\frac{1}{4}}y\right)\mu_{\Gamma_k},$$

i.e. a kind of weighted density measure. Clearly,

$$\int_{\Gamma_k} f\varphi_1 \, d\mathcal{H}^1 = \fint_{\Gamma_k} D\varphi \, d\mu_k^0 + \tfrac{1}{k}\varphi_1(x_k, y_k)$$

for all $\varphi \in \mathcal{W}^{1,\infty}(\Omega, \mathbb{R}^2)$ and odd k. Analogously we get measures μ_k^0 for even k. Since the μ_k^0 have the same total variation as the $G^0\mathcal{H}^1\lfloor\Gamma_k$, their sum is again a measure and we can replace the Radon measures $G^0\mathcal{H}^1\lfloor\Gamma_k$ with the μ_k^0 to handle $\varphi \in \mathcal{W}^{1,\infty}(\Omega, \mathbb{R}^2)$. By analogous arguments we can construct measures $\mu_k^1 \in \mathrm{ba}(\Omega, \mathcal{B}(\Omega), \mathcal{L}^2)^{2\times 2}$ with core Γ_k^1 that can replace the Radon measures $G_k^1\mathcal{H}^1\lfloor\Gamma_k^1$. This way we can finally replace the former measure μ_f with

$$\mu_f := \sum_{k=1}^{\infty} \mu_{2k+1}^0 - \mu_{2k}^0 - \mu_k^1$$

to get a Gauss-Green formula for all $\varphi \in \mathcal{W}^{1,\infty}(\Omega, \mathbb{R}^2)$.

Notice that for the derivation of the Gauss-Green formula we did not use (4.90). Since σ_f and μ_f do not depend on δ, we could thus apply Proposition 4.41 after this derivation to get (4.90). Recall that the choice of σ_f and μ_f is not unique, since an alternative version with merely terms for $k > k_0$ in μ_f is possible for any $k_0 \in \mathbb{N}$.

Let us now discuss normal measures for Sobolev and BV functions. We start with a scalar variant of Proposition 4.25.

Proposition 4.46 *Let $U \subset \mathbb{R}^n$ be open, bounded, let $\Omega \in \mathcal{B}(U)$, let $f \in \mathcal{L}^1(\Omega)$, let ν be a normal measure related to a good approximation χ for χ_Ω and with approximating sequence $\{\chi_k\}$ satisfying $|D\chi_k| \le \gamma k \ \mathcal{L}^n$-a.e. on Ω for some $\gamma > 0$, and let $A \subset U$ be an aura of ν as in* (4.32). *If there is some $\tilde{\delta} > 0$ such that*

$$\frac{1}{\delta}\int_{(\partial\Omega)_\delta \cap A} |f| \, d\mathcal{L}^n \quad \text{is uniformly bounded for} \quad 0 < \delta < \tilde{\delta}, \tag{4.101}$$

then φf is ν-integrable on U for all $\varphi \in \mathcal{L}^\infty(U, \mathbb{R}^n)$. If we have in addition that

$$\lim_{k\to\infty} \frac{1}{\delta}\int_{(\partial\Omega)_\delta \cap A \cap \{|f| \ge k\}} |f| \, d\mathcal{L}^n = 0 \quad \text{uniformly for } \delta \in (0, \tilde{\delta}), \tag{4.102}$$

then for each $\varphi \in \mathcal{L}^\infty(U, \mathbb{R}^n)$ *there is a subsequence* $\{\chi_{k'}\}$ *such that*

$$\lim_{k'\to\infty} \int_U f\varphi \cdot D\chi_{k'}\, d\mathcal{L}^n = -\!\!\!\!\!\!\int_{\partial\Omega} f\varphi\, d\nu\,. \tag{4.103}$$

For the proof we apply Proposition 4.25 to vector fields F_k as in the proof of Theorem 4.40. Analogously we obtain the subsequent results from Theorem 4.27, Corollary 4.29 and Proposition 4.30.

Theorem 4.47 *Let* $U \subset \mathbb{R}^n$ *be open and bounded, let* $\Omega \in \mathcal{B}(U)$*, let* ν *be a normal measure of* Ω *related to a good approximation* χ *for* χ_Ω *with approximating sequence* $\{\chi_k\}$*, let* $f \in \mathcal{BV}(U)$ *be* ν*-integrable such that* (4.103) *is satisfied, and let* $\chi_k \to \chi$ Df*-a.e. on* $\partial\Omega$*. Then we have for all functions* $\varphi \in \mathcal{W}^{1,\infty}(U, \mathbb{R}^n)$ *that*

$$\int_{\partial\Omega} \chi\, d\,\mathrm{div}(f\varphi) + \mathrm{div}(f\varphi)(\mathrm{int}\,\Omega) = -\!\!\!\!\!\!\int_{\partial\Omega} f\varphi\, d\nu\,. \tag{4.104}$$

If $f \in \mathcal{BV}(U) \cap \mathcal{L}^\infty(U)$*, then* f *is* ν*-integrable, satisfies* (4.103)*, and we have* (4.104) *for all* $\varphi \in \mathcal{W}^{1,\infty}(U, \mathbb{R}^n)$.

Proposition 4.48 *Let* $U \subset \mathbb{R}^n$ *be open and bounded, let* $\Omega \in \mathcal{B}(U)$*, let* ν^* *be a normal measure where* $*$ *stands for* int, intc, ext, *or* sym*, let* χ^**,* χ_k^**, and* δ_k *be related to* ν^* *as in the corresponding examples above, let* $\omega_{\partial\Omega}^* \in \mathrm{ba}\big(U, \mathcal{B}(U), \mathcal{L}^n\big)$ *be the related measure as in Proposition 4.30, and let* ν^Ω *be the normal field from* (4.2)*. If* $f \in \mathcal{BV}(U)$ *is* ν^**-integrable such that* (4.103) *is satisfied and that* $\chi_k^* \to \chi^*$ Df*-a.e. on* $\partial\Omega$*, then we have for all* $\varphi \in \mathcal{W}^{1,\infty}(U, \mathbb{R}^n)$

$$\mathrm{div}(f\varphi)(\mathrm{int}\,\Omega) = -\!\!\!\!\!\!\int_{\partial\Omega} f\varphi \cdot \nu^\Omega\, d\omega_{\partial\Omega}^{\mathrm{int}}\,,$$

$$\mathrm{div}(f\varphi)(\mathrm{int}\,\Omega) = -\!\!\!\!\!\!\int_{\partial\Omega} f\varphi \cdot \nu^\Omega\, d\omega_{\partial\Omega}^{\mathrm{intc}}\,,$$

$$\mathrm{div}(f\varphi)(\overline{\Omega}) = -\!\!\!\!\!\!\int_{\partial\Omega} f\varphi \cdot \nu^\Omega\, d\omega_{\partial\Omega}^{\mathrm{ext}}\,,$$

$$\tfrac{1}{2}\,\mathrm{div}(f\varphi)(\partial\Omega) + \mathrm{div}(f\varphi)(\mathrm{int}\,\Omega) = -\!\!\!\!\!\!\int_{\partial\Omega} f\varphi \cdot \nu^\Omega\, d\omega_{\partial\Omega}^{\mathrm{sym}}\,.$$

For bounded functions on sets of finite perimeter we can transfer Proposition 4.34 (cf. also [12, p. 259] and use [2, pp. 171, 177] for (4.111)).

Proposition 4.49 *Let* $U \subset \mathbb{R}^n$ *be open and bounded, let* $\Omega \subset U$ *have finite perimeter, let* $f \in \mathcal{BV}(U) \cap \mathcal{L}^\infty(U)$*, and let* $\tilde{\nu}^{\mathrm{int}}$*,* $\tilde{\nu}^{\mathrm{ext}}$ *be the normal measures from Examples 4.21 and 4.22.*

(1) *If* $\Omega \Subset U$*, then there are vector-valued functions*

$$F^{\mathrm{int}}, F^{\mathrm{ext}} \in \mathcal{L}^\infty(\partial_*\Omega, \mathcal{H}^{n-1})^n \quad \textit{with} \quad \|F^{\mathrm{int}}\|_\infty,\ \|F^{\mathrm{ext}}\|_\infty \le \|f\|_{\partial\Omega}$$

such that

$$\operatorname{div}(\varphi f)(\operatorname{int}_*\Omega) = \fint_{\partial\Omega} f\varphi\, d\tilde{\nu}^{\text{int}} = \int_{\partial_*\Omega} \varphi F^{\text{int}}\, d\mathcal{H}^{n-1}, \tag{4.105}$$

$$\operatorname{div}(\varphi f)(\partial_*\Omega \cup \operatorname{int}_*\Omega) = \fint_{\partial\Omega} f\varphi\, d\tilde{\nu}^{\text{ext}} = \int_{\partial_*\Omega} \varphi F^{\text{ext}}\, d\mathcal{H}^{n-1} \tag{4.106}$$

for all $\varphi \in \mathcal{W}^{1,\infty}(U, \mathbb{R}^n)$ *(cf.* (2.10) *for* $\|\cdot\|_{\partial\Omega}$*). If* Ω *is open we also have*

$$\operatorname{div}(\varphi F)(\Omega) = \int_{\partial_*\Omega} \varphi F^{\text{int}}\, d\mathcal{H}^{n-1} - \int_{\operatorname{int}_*\Omega\cap\partial\Omega} \varphi\, dDf \tag{4.107}$$

and if Ω *is closed*

$$\operatorname{div}(\varphi F)(\Omega) = \int_{\partial_*\Omega} \varphi F^{\text{ext}}\, d\mathcal{H}^{n-1} + \int_{\operatorname{ext}_*\Omega\cap\partial\Omega} \varphi\, dDf \tag{4.108}$$

for all $\varphi \in \mathcal{W}^{1,\infty}(U, \mathbb{R}^n)$.

(2) *Let* $\Omega \subset U$ *be open with* $\mathcal{H}^{n-1}(\partial\Omega \cap \operatorname{int}_*\Omega) < \infty$. *Then we have that* $\hat{f} \in \mathcal{BV}(\mathbb{R}^n) \cap \mathcal{L}^\infty(\mathbb{R}^n)$ *for the extension* $\hat{f}$ *of* f *by zero outside of* Ω. *There is also some normal measure* $\nu \in \operatorname{ba}\big(U, \mathcal{B}(U), \mathcal{L}^n\big)^n$ *such that* $(\partial\Omega)_\delta \cap \Omega$ *is an aura for all* $\delta > 0$ *and*

$$\operatorname{div}(\varphi f)(\Omega) = \fint_{\partial\Omega} f\varphi\, d\nu \tag{4.109}$$

for all $\varphi \in \mathcal{W}^{1,\infty}(U, \mathbb{R}^n)$. *Moreover there is* $F \in \mathcal{L}^\infty\big(\partial\Omega \setminus \operatorname{ext}_*\Omega, \mathcal{H}^{n-1}\big)^n$ *with* $\|F\|_\infty \le c\|f\|_{\partial\Omega}$ *for some* $c > 0$ *depending on dimension* n *such that*

$$\operatorname{div}(\varphi f)(\Omega) = \int_{\partial\Omega\setminus\operatorname{ext}_*\Omega} \varphi F\, d\mathcal{H}^{n-1} \tag{4.110}$$

for all $\varphi \in C^1(\mathbb{R}^n, \mathbb{R}^n)$ *where*

$$F\mathcal{H}^{n-1}\llcorner(\partial\Omega \cap \operatorname{int}_*\Omega) = -Df\llcorner(\partial\Omega \cap \operatorname{int}_*\Omega) \qquad \textit{and}$$

$$F\mathcal{H}^{n-1}\llcorner(\partial_*\Omega) = \tilde{f}\nu^\Omega\mathcal{H}^{n-1}\llcorner(\partial_*\Omega) \qquad \textit{with}$$

$$\tilde{f}(x) = \lim_{r\downarrow 0} \fint_{B_r(x)\cap\Omega} f\, d\mathcal{L}^n \quad \mathcal{H}^{n-1}\text{-a.e. on } \partial_*\Omega\,. \tag{4.111}$$

If $\hat{f}$ is continuous on a neighborhood of Ω, then $(Df)(\partial\Omega \cap \mathrm{int}_*\Omega) = 0$ *and*

$$\mathrm{div}(\varphi f)(\mathrm{int}_*\Omega) = \mathrm{div}(\varphi f)(\Omega) = \int_{\partial_*\Omega} f\varphi \cdot \nu^\Omega \, d\mathcal{H}^{n-1} \tag{4.112}$$

for all $\varphi \in C^1(\mathbb{R}^n, \mathbb{R}^n)$.

Let us now consider the case where $U = \Omega \subset \mathbb{R}^n$ has Lipschitz boundary and let $f \in \mathcal{BV}(\Omega)$. By Corollary 3.18 we have case (C) for $\Gamma = \partial\Omega$ and all $\delta > 0$ and by Proposition 4.44 we have (4.90). Hence Theorem 4.40 implies

$$\mathrm{div}(\varphi f)(\Omega) = \int_{\partial\Omega} \varphi \, d\sigma_f + \fint_{\partial\Omega} D\varphi \, d\mu_f \quad \text{for all } \varphi \in \mathcal{W}^{1,\infty}(\Omega, \mathbb{R}^n) \tag{4.113}$$

where σ_f is a Radon measure supported on $\partial\Omega$ and core $\mu_f \subset \partial\Omega$. With the precise representative $f^\times$ according to Remark 2.16, we get from the literature that

$$\mathrm{div}(\varphi f)(\Omega) = \int_{\partial\Omega} f^\times \varphi \cdot \nu \, d\mathcal{H}^{n-1} \quad \text{for all} \quad \varphi \in \mathcal{W}^{1,\infty}(\Omega, \mathbb{R}^n) \tag{4.114}$$

(cf. [32, p. 168], [22, p. 177]). Therefore we can choose $\mu_f = 0$ in (4.113) by Proposition 4.41(3). In this case we have

$$\sigma_f = f^\times \nu \mathcal{H}^{n-1} \llcorner \partial\Omega$$

since all $\varphi \in C(\overline{\Omega}, \mathbb{R}^n)$ can be uniformly approximated by functions from the space $\mathcal{W}^{1,\infty}(\Omega, \mathbb{R}^n)$. Notice that this version of the Gauss-Green formula with a σ-measure σ_f supported on the boundary of Ω requires a pointwise trace function $f^\times$ on $\partial\Omega$. Let us now provide an alternative version where only the values of f on Ω are used. For that we verify (4.104) with $\chi = \chi_{\mathrm{int}\,\Omega}$ and $\nu = \nu^{\mathrm{int}}$.

Theorem 4.50 *Let $U = \Omega \subset \mathbb{R}^n$ be open and bounded with Lipschitz boundary, let $f \in \mathcal{BV}(\Omega)$, let ν^{int} be the interior normal measure from Example 4.21, let $\omega^{\mathrm{int}}_{\partial\Omega}$ be the measure according to Proposition 4.30, and let ν^Ω be the normal field as in* (4.2). *Then f is ν^{int}-integrable and*

$$\begin{aligned}\mathrm{div}(\varphi f)(\Omega) &= \int_\Omega f \, \mathrm{div}\,\varphi \, d\mathcal{L}^n + \int_\Omega \varphi \, dDf \\ &= \fint_{\partial\Omega} f\varphi \, d\nu^{\mathrm{int}} = \fint_{\partial\Omega} f\varphi \cdot \nu^\Omega \, d\omega^{\mathrm{int}}_{\partial\Omega}\end{aligned} \tag{4.115}$$

for all $\varphi \in \mathcal{W}^{1,\infty}(\Omega, \mathbb{R}^n)$.

By $\mathcal{W}^{1,1}(\Omega) \subset \mathcal{BV}(\Omega)$ we have that (4.115) is also valid for all Sobolev functions f (cf. Remark 3.8). For the proof, that is given after the proof of the next lemma, we do not directly apply Theorem 4.47 that is based on the technical condition (4.103). We rather show the assertion directly by using an approximating sequence of f. Nevertheless, for the approximating sequence χ_k^{int} of $\chi_{\text{int}\,\Omega}$ related to ν^{int} according to Example 4.21, we readily get (4.103) at the end of the proof. Notice that we can use χ_k^{int} from (4.50), since $\Omega = \text{int}\,\Omega$ and thus, $\text{div}(\varphi f)(\Omega) = 0$ if merely $\varphi = 0$ on $\partial\Omega$ (cf. Proposition 3.7). Alternatively we can use χ_k^{intc} from Example 4.21 that might give a slightly different normal measure ν^{intc}, but the related integral in (4.115) would give the same values (roughly speaking, the integral performs a slightly different averaging near $\partial\Omega$ that does not change the result for functions entering (4.115); cf. also Remark 4.28 (2)). In the proof we use arguments that are similar to those in the usual proof about traces (cf. [22, pp. 177–181]), however we have to work them out in much more detail. But, let us first formulate a simple consequence of the theorem.

Corollary 4.51 *Let $U \subset \mathbb{R}^n$ be open and bounded, let $\Omega \Subset U$ be open with Lipschitz boundary, let $f \in \mathcal{BV}(U)$, let ν^{ext} be the exterior normal measure from Example 4.22, let $\omega_{\partial\Omega}^{\text{ext}}$ be as in Proposition 4.30, and let the normal field ν^Ω be as in (4.2). Then f is ν^{ext}-integrable and*

$$\begin{aligned} \text{div}(\varphi f)(\overline{\Omega}) &= \int_{\overline{\Omega}} f \,\text{div}\,\varphi \, d\mathcal{L}^n + \int_{\overline{\Omega}} \varphi \, dDf \\ &= \oint_{\partial\Omega} f\varphi \, d\nu^{\text{ext}} = \oint_{\partial\Omega} f\varphi \cdot \nu^\Omega \, d\omega_{\partial\Omega}^{\text{ext}} \end{aligned} \tag{4.116}$$

for all $\varphi \in \mathcal{W}^{1,\infty}(\Omega, \mathbb{R}^n)$.

For the use of ν^{ext} the function f has to be given in a small neighborhood of Ω. The results in (4.115) and (4.116) will differ if $|Df|(\partial\Omega) \neq 0$. But we have $|Df|(\partial\Omega) = 0$ for Sobolev functions and thus, both formulas give the same in that case.

Proof Let Ω_0 be open with Lipschitz boundary such that $\Omega \Subset \Omega_0 \Subset U$. We change f to be zero on $U \setminus \Omega_0$ and still have $f \in \mathcal{BV}(U)$ (cf. [22, p. 183]). Moreover we can assume that U has Lipschitz boundary. From (4.115) we get

$$\text{div}(\varphi f)(U) = 0\,, \quad \text{div}(\varphi f)(U \setminus \overline{\Omega}) = \oint_{\partial\Omega} f\varphi \, d\nu_{\Omega^c}^{\text{int}}$$

where $\nu_{\Omega^c}^{\text{int}}$ is the interior normal measure of Ω^c (cf. Example 4.21). By construction we readily see that $\nu_{\Omega^c}^{\text{int}} = -\nu^{\text{ext}}$. Consequently

$$\text{div}(\varphi f)(\overline{\Omega}) = \text{div}(\varphi f)(U) - \text{div}(\varphi f)(U \setminus \overline{\Omega}) = \oint_{\partial\Omega} f\varphi \, d\nu^{\text{ext}}$$

which gives the first assertion. For the second one we use that $\omega_{\partial\Omega}^{\text{ext}} = \omega_{\partial(\Omega^c)}^{\text{int}}$ by (4.63) and that $\nu^\Omega = -\nu^{\Omega^c}$ by (4.2). □

Proof of Theorem 4.50 Let us fix some $\varphi \in \mathcal{W}^{1,\infty}(\Omega, \mathbb{R}^n)$ and recall Example 4.21. Since Ω has Lipschitz boundary, we have (4.49) for any sequence $\delta_m \le \frac{1}{m}$. For the approximating sequence $\chi_m = \chi_m^{\mathrm{int}}$ of $\chi_{\mathrm{int}\,\Omega}$ according to (4.50), we set $\psi_m := 1 - \chi_m$. Notice that $\psi_m \varphi \in \mathcal{W}^{1,\infty}(\Omega, \mathbb{R}^n)$ for all m. Since $\Omega = \mathrm{int}\,\Omega$ and since ψ_m equals 1 on $\partial\Omega$, we get from Proposition 3.7 for all $g \in \mathcal{BV}(\Omega)$ and all $m \in \mathbb{N}$ (cf. also [2, p. 118])

$$
\begin{aligned}
&\operatorname{div}(\varphi g)(\Omega) \\
&\quad = \operatorname{div}(\psi_m \varphi g)(\Omega) \\
&\quad = \int_\Omega g \operatorname{div}(\psi_m \varphi)\, d\mathcal{L}^n + \int_\Omega \psi_m \varphi\, dDg \\
&\quad = \int_\Omega g \psi_m \operatorname{div} \varphi\, d\mathcal{L}^n + \int_\Omega g \varphi D\psi_m\, d\mathcal{L}^n + \int_\Omega \psi_m \varphi\, dDg\,. \qquad (4.117)
\end{aligned}
$$

The first and the last integral in (4.117) tend to zero for $m \to \infty$ and thus,

$$
\lim_{m\to\infty} \int_\Omega g\varphi \cdot D\psi_m\, d\mathcal{L}^n = \operatorname{div}(\varphi g)(\Omega)\,. \qquad (4.118)
$$

Choose now an approximating sequence $f_k \in \mathcal{BV}(\Omega) \cap C^\infty(\Omega)$ for $f \in \mathcal{BV}(\Omega)$ with

$$
f_k \to f \ \text{ in } \mathcal{L}^1(\Omega)\,, \quad |Df_k|(\Omega) \to |Df|(\Omega)\,, \quad Df_k \overset{*}{\rightharpoonup} Df \qquad (4.119)
$$

where the last convergence denotes the weak* convergence in the sense of Radon measures with $(Df_k)(B) = \int_B Df_k\, d\mathcal{L}^n$ for $B \in \mathcal{B}(\Omega)$ (cf. [22, pp. 54, 172, 175]). Then

$$
\begin{aligned}
\lim_{k\to\infty} \operatorname{div}(\varphi f_k)(\Omega) &= \lim_{k\to\infty} \left(\int_\Omega f_k \operatorname{div} \varphi\, d\mathcal{L}^n + \int_\Omega \varphi\, dDf_k \right) \\
&= \operatorname{div}(\varphi f)(\Omega)\,.
\end{aligned}
$$

By Lemmas 4.52 and 4.53 below we have that f is ν^{int}-integrable, that

$$
\lim_{k\to\infty} \fint_{\partial\Omega} f_k \varphi\, d\nu^{\mathrm{int}} = \fint_{\partial\Omega} f\varphi\, d\nu^{\mathrm{int}}\,, \qquad (4.120)
$$

and that

$$
\lim_{m\to\infty} \int_\Omega f_k \varphi \cdot D\psi_m\, d\mathcal{L}^n = \fint_{\partial\Omega} f_k \varphi\, d\nu^{\mathrm{int}} \quad \text{for all } k \in \mathbb{N} \qquad (4.121)
$$

(notice that Proposition 4.46 implies (4.121) merely for a subsequence of $\{\psi_m\}$ that depends on f_k). From (4.118) for $g = f_k$ and from (4.121) we get

$$\operatorname{div}(\varphi f_k)(\Omega) = \fint_{\partial\Omega} f_k \varphi \, d\nu^{\text{int}} \quad \text{for all } k \in \mathbb{N}\,.$$

Consequently

$$\begin{aligned} &\Big| \operatorname{div}(\varphi f)(\Omega) - \fint_{\partial\Omega} f\varphi \, d\nu^{\text{int}} \Big| \\ &\quad \le \big| \operatorname{div}(\varphi f)(\Omega) - \operatorname{div}(\varphi f_k)(\Omega) \big| + \Big| \fint_{\partial\Omega} f_k \varphi \, d\nu^{\text{int}} - \fint_{\partial\Omega} f\varphi \, d\nu^{\text{int}} \Big|\,. \end{aligned}$$

Since the right hand side tends to zero as $k \to \infty$, we get

$$\operatorname{div}(\varphi f)(\Omega) = \fint_{\partial\Omega} f\varphi \, d\nu^{\text{int}}$$

which verifies the first equality in (4.115). From (4.118) we get

$$\lim_{m\to\infty} \int_\Omega f\varphi \cdot D\psi_m \, d\mathcal{L}^n = \fint_{\partial\Omega} f\varphi \, d\nu^{\text{int}}\,.$$

Thus we can apply Proposition 4.48 to get the second equality in (4.115). □

Lemma 4.52 *Let $\Omega \subset \mathbb{R}^n$ be open and bounded with Lipschitz boundary and let $g \in \mathcal{BV}(\Omega) \cap C^\infty(\Omega)$. Moreover let ν^{int} be the interior normal measure of Ω with approximating sequence χ_m^{int} according to* (4.50). *Then*

$$\lim_{m\to\infty} \int_\Omega g\varphi \cdot D\chi_m^{\text{int}} \, d\mathcal{L}^n = -\fint_{\partial\Omega} g\varphi \, d\nu^{\text{int}} \tag{4.122}$$

for all $\varphi \in \mathcal{W}^{1,\infty}(\Omega, \mathbb{R}^n)$.

Proof For $x \in \mathbb{R}^n$ we use the notation

$$x = (x_1, \ldots, x_n) = (x', x_n) \in \mathbb{R}^n \quad \text{with} \quad x' = (x_1, \ldots, x_{n-1}) \in \mathbb{R}^{n-1}\,.$$

Since Ω has Lipschitz boundary, for each $x \in \partial\Omega$ there is a cylinder

$$C(x, r, h) := \{(y', y_n) \mid |y' - x'| < r\,,\ |y_n - x_n| < 2h\}$$

and a Lipschitz continuous function γ on $B_r(x') \subset \mathbb{R}^{n-1}$ such that after a suitable rotation of the coordinate system $|\gamma(y') - x_n| < h$ on $B_r(x')$ and

$$\Omega \cap C(x, r, h) = \{y \in \mathbb{R}^n \mid |y' - x'| < r\,,\ \gamma(y') < y_n < x_n + 2h\}\,.$$

Since $\partial\Omega$ can be covered by finitely many such cylinders, it is sufficient to show (4.122) only for the case where the integrals are restricted to a cylinder $C := C(\bar{x}, r, h)$ and to work in the related coordinate system. The general case then follows by a straightforward argument with a partition of unity subordinate to finitely many such cylinders.

Let us fix some $\varphi \in \mathcal{W}^{1,\infty}(\Omega, \mathbb{R}^n)$ and a cylinder $C = C(\bar{x}, r, h)$ with $\bar{x} \in \partial\Omega$ and let us define

$$C^{s,t} := \{x \in C \mid \gamma(x') + s < x_n < \gamma(x') + t\} \quad \text{for} \quad 0 \le s < t < h\,,$$

$$g^t(x) := g(x', \gamma(x') + t)\,, \quad \tilde{g}^t(x) := g(x', x_n + t) \quad \text{for} \quad 0 < t < h\,.$$

$\tilde{g}^t$ is a shift of g with $\tilde{g}^t \in C^\infty(\overline{\Omega \cap C})$, while g^t is constant in the last coordinate and not necessarily smooth. Moreover we briefly write $\chi_m = \chi_m^{\text{int}}$ and let $\delta_m > 0$ be related to it according to (4.50). Since $\partial\Omega$ is Lipschitz, there is some $\tilde{c} > 0$ such that

$$(\operatorname{supp} D\chi_m) \cap C = (\partial\Omega)_{\delta_m} \cap \Omega \cap C \subset C^{0,\beta_m} \quad \text{for} \quad \beta_m := \tilde{c}\delta_m\,. \tag{4.123}$$

(a) For $0 < s < t < h$ we now have

$$\begin{aligned} |g^t(x) - g^s(x)| &\le \int_s^t \left| \frac{\partial g}{\partial x_n}(x', \gamma(x') + \tau) \right| d\tau \\ &\le \int_s^t |Dg(x', \gamma(x') + \tau)|\, d\tau\,. \end{aligned}$$

By the coarea formula,

$$\int_{\partial\Omega\cap C} |g^t - g^s|\, d\mathcal{H}^{n-1} \le \int_{C^{s,t}} |Dg|\, d\mathcal{L}^n \le \int_{C^{0,t}} |Dg|\, d\mathcal{L}^n\,. \tag{4.124}$$

The right hand side tends to zero as $t \to 0$ and thus, there is some g^0 such that

$$\lim_{t\to 0} g^t = g^0 \quad \text{in} \quad \mathcal{L}^1(\partial\Omega \cap C, \mathcal{H}^{n-1})\,. \tag{4.125}$$

We can extend g^0 on C such that $g^0(x) = g^0(x', \gamma(x'))$. By (4.124) and the integrability of $|Dg|$, for any $\varepsilon > 0$ there is some $\delta > 0$ such that

$$\int_{\partial\Omega\cap C} |g^t - g^s|\, d\mathcal{H}^{n-1} < \varepsilon \quad \text{whenever} \quad |t - s| < \delta \tag{4.126}$$

(cf. [53, p. 1016]). Below we show for $t > 0$ that

$$\lim_{m\to\infty} \int_{\Omega\cap C} \tilde{g}^t \varphi \cdot D\chi_m \, d\mathcal{L}^n = -\int_{\partial\Omega\cap C} g^t \varphi \cdot \nu^{\Omega} \, d\mathcal{H}^{n-1} , \tag{4.127}$$

$$\fint_{\partial\Omega\cap C} g^t \varphi \, d\nu^{\text{int}} = \int_{\partial\Omega\cap C} g^t \varphi \cdot \nu^{\Omega} \, d\mathcal{H}^{n-1} , \tag{4.128}$$

$$\lim_{t\to 0} \int_{\Omega\cap C} \tilde{g}^t \varphi \cdot D\chi_m \, d\mathcal{L}^n = \int_{\Omega\cap C} g\varphi \cdot D\chi_m \, d\mathcal{L}^n \quad \text{uniformly for } m \in \mathbb{N} , \tag{4.129}$$

$$\lim_{t\to 0} \fint_{\partial\Omega\cap C} g^t \varphi \, d\nu^{\text{int}} = \fint_{\partial\Omega\cap C} g\varphi \, d\nu^{\text{int}} . \tag{4.130}$$

Consequently, for $\varepsilon > 0$ there is some $t_0 > 0$ and some $m_0 \in \mathbb{N}$ such that we have

$$\begin{aligned}
&\Big| \int_{\Omega\cap C} g\varphi \cdot D\chi_m \, d\mathcal{L}^n + \fint_{\partial\Omega\cap C} g\varphi \, d\nu^{\text{int}} \Big| \\
&\le \Big| \int_{\Omega\cap C} g\varphi \cdot D\chi_m \, d\mathcal{L}^n - \int_{\Omega\cap C} \tilde{g}^{t_0} \varphi \cdot D\chi_m \, d\mathcal{L}^n \Big| \\
&\quad + \Big| \int_{\Omega\cap C} \tilde{g}^{t_0} \varphi \cdot D\chi_m \, d\mathcal{L}^n + \int_{\partial\Omega\cap C} g^{t_0} \varphi \cdot \nu^{\Omega} \, d\mathcal{H}^{n-1} \Big| \\
&\quad + \Big| \fint_{\partial\Omega\cap C} g\varphi \, d\nu^{\text{int}} - \fint_{\partial\Omega\cap C} g^{t_0} \varphi \, d\nu^{\text{int}} \Big| \\
&\le 3\varepsilon \qquad \text{for all } m > m_0 .
\end{aligned}$$

But this implies the assertion (4.122) and it remains to show (4.127), (4.128), (4.129), and (4.130).

(b) Let us show (4.127). We have that $\tilde{g}^t, \chi_m \in \mathcal{W}^{1,\infty}(\Omega)$ and that $\Omega \cap C$ has Lipschitz boundary. Thus, integration by parts gives for $t > 0$

$$\begin{aligned}
&-\int_{\Omega\cap C} \tilde{g}^t \varphi \cdot D\chi_m \, d\mathcal{L}^n \\
&= \int_{\Omega\cap C} \tilde{g}^t \varphi \cdot D(1-\chi_m) \, d\mathcal{L}^n \\
&= -\int_{\Omega\cap C} (1-\chi_m) \operatorname{div}(\tilde{g}^t \varphi) \, d\mathcal{L}^n
\end{aligned}$$

$$
\begin{aligned}
&+\int_{\partial(\Omega\cap C)} (1-\chi_m)\,\tilde{g}^t\varphi\cdot\nu^{\Omega\cap C}\,d\mathcal{H}^{n-1} \\
&\overset{m\to\infty}{\longrightarrow} \int_{\partial\Omega\cap C} \tilde{g}^t\varphi\cdot\nu^{\Omega}\,d\mathcal{H}^{n-1} = \int_{\partial\Omega\cap C} g^t\varphi\cdot\nu^{\Omega}\,d\mathcal{H}^{n-1}\,.
\end{aligned}
$$

But this is (4.127).

(c) For (4.128) use $\tilde{g}^t, g^t \in C(\overline{\Omega\cap C})$ and $\|g^t-\tilde{g}^t\|_{C((\partial\Omega)_\delta\cap\Omega\cap C)} \overset{\delta\downarrow 0}{\longrightarrow} 0$. Then, by Theorem 4.19, there is a subsequence $\{\chi_{m'}\}$ with

$$
\lim_{m'\to\infty}\int_{\Omega\cap C}\tilde{g}^t\varphi\cdot D\chi_{m'}\,d\mathcal{L}^n = -\fint_{\partial\Omega\cap C}\tilde{g}^t\varphi\,d\nu^{\text{int}} = -\fint_{\partial\Omega\cap C} g^t\varphi\,d\nu^{\text{int}}\,.
$$

With (4.127) we get (4.128).

(d) Let us verify (4.129). For $\varepsilon>0$ we choose $\delta>0$ as in (4.126) and obtain for all $m\in\mathbb{N}$

$$
\begin{aligned}
&\int_{\Omega\cap C}|\tilde{g}^t-g|\,\varphi\cdot D\chi_m\,d\mathcal{L}^n \\
&\le \frac{\|\varphi\|_\infty}{\delta_m}\int_{C^{0,\beta_m}}|\tilde{g}^t-g|\,d\mathcal{L}^n \\
&= \frac{\|\varphi\|_\infty}{\delta_m}\int_0^{\beta_m}\int_{\partial\Omega\cap C}|g^{s+t}-g^s|\,d\mathcal{H}^{n-1}\,ds \\
&\le \frac{\|\varphi\|_\infty\beta_m\varepsilon}{\delta_m} = \|\varphi\|_\infty\,\tilde{c}\,\varepsilon \qquad \text{for all } 0<t<\delta,\ m\in\mathbb{N}\,.
\end{aligned}
$$

This gives (4.129).

(e) As preparation for the proof of (4.130) we first show that $g^t \overset{\nu^{\text{int}}}{\longrightarrow} g$. Let us fix some $\varepsilon>0$ and notice that $g^t \overset{\mathcal{H}^{n-1}}{\longrightarrow} g^0$ on $\partial\Omega$ by (4.125). Therefore we have

$$
\mathcal{H}^{n-1}(B^t)\overset{t\to 0}{\longrightarrow} 0 \quad\text{where}\quad B^t := \big\{x\in\partial\Omega\cap C \,\big|\, |g^t-g^0|>\varepsilon\big\}\,. \tag{4.131}
$$

Obviously, for $s,t>0$,

$$
\begin{aligned}
B^{t,s} &:= \big\{x\in\partial\Omega\cap C \,\big|\, |g^t-g^s|>2\varepsilon\big\} \\
&\subset \big\{x\in\partial\Omega\cap C \,\big|\, |g^t-g^0|+|g^s-g^0|>2\varepsilon\big\} \subset B^t\cup B^s\,.
\end{aligned}
$$

For $\delta>0$ and

$$
\tilde{B}^t := \big\{x\in\Omega\cap C \,\big|\, |g^t-g|>2\varepsilon\big\}\,,
$$

we have that

$$\mathcal{L}^n(\tilde{B}^t \cap C^{0,\delta}) = \int_0^\delta \mathcal{H}^{n-1}(B^{t,s})\,ds \le \int_0^\delta \mathcal{H}^{n-1}(B^t) + \mathcal{H}^{n-1}(B^s)\,ds\,.$$

By Theorem 4.19,

$$\begin{aligned}
|\nu^{\text{int}}|(\tilde{B}^t) &\le \limsup_{m\to\infty} \|D\chi_m\|_{\mathcal{L}^1(\tilde{B}^t)} = \limsup_{m\to\infty} \int_{\tilde{B}^t} |D\chi_m|\,d\mathcal{L}^n \\
&\le \limsup_{m\to\infty} \tfrac{1}{\delta_m} \int_{\tilde{B}^t\cap C^{0,\beta_m}} d\mathcal{L}^n \\
&\le \limsup_{m\to\infty} \tfrac{1}{\delta_m} \int_0^{\beta_m} \mathcal{H}^{n-1}(B^t) + \mathcal{H}^{n-1}(B^s)\,ds \\
&\le \limsup_{m\to\infty} \tfrac{2\beta_m}{\delta_m} \sup_{s\in(0,t]} \mathcal{H}^{n-1}(B^s) \quad (\text{use } \beta_m < t \text{ for } m \text{ large}) \\
&= 2\tilde{c} \sup_{s\in(0,t]} \mathcal{H}^{n-1}(B^s)\,.
\end{aligned}$$

By (4.131) the right hand side tends to zero as $t \to 0$. Since $\varepsilon > 0$ was arbitrary, we obtain $g^t \xrightarrow{\nu^{\text{int}}} g$. For φg^t we have

$$\tilde{B}^t_\varphi := \big\{x \in \Omega\cap C \;\big|\; |\varphi g^t - \varphi g| > \varepsilon\big\} \subset \big\{x \in \Omega\cap C \;\big|\; \|\varphi\|_\infty |g^t - g| > \varepsilon\big\}\,.$$

Hence, $|\nu^{\text{int}}|(\tilde{B}^t) \to 0$ for all $\varepsilon > 0$ implies that $|\nu^{\text{int}}|(\tilde{B}^t_\varphi) \to 0$ for all $\varepsilon > 0$. Thus we also have $\varphi g^t \xrightarrow{\nu^{\text{int}}} \varphi g$.

(f) Finally we prove (4.130). Let us fix some $\varepsilon > 0$. For fixed s, t we consider the measure $\nu := \|\varphi\|_\infty |g^t - g^s| |\nu^{\text{int}}|$. By (2.9) there is some $\tilde{\varphi} \in \mathcal{L}^\infty(\Omega, \mathbb{R}^n)$ with $\|\tilde{\varphi}\|_\infty \le 1$ and by Theorem 4.19 there is some subsequence $\chi_{m'}$ such that

$$\begin{aligned}
&\int_{\Omega\cap C} \|\varphi\|_\infty |g^t - g^s|\,d|\nu^{\text{int}}| - \varepsilon \\
&\quad\le \|\varphi\|_\infty \int_{\Omega\cap C} |g^t - g^s| \tilde{\varphi}\,d\nu^{\text{int}} \\
&\quad= \lim_{m'\to\infty} \|\varphi\|_\infty \int_{\Omega\cap C} |g^t - g^s| \tilde{\varphi}\cdot D\chi_{m'}\,d\mathcal{L}^n \\
&\quad\le \limsup_{m\to\infty} \frac{\|\varphi\|_\infty}{\delta_m} \int_{(\partial\Omega)_{\delta_m}\cap\Omega\cap C} |g^t - g^s|\,d\mathcal{L}^n \\
&\quad\le \limsup_{m\to\infty} \frac{\|\varphi\|_\infty}{\delta_m} \int_{C^{0,\beta_m}} |g^t - g^s|\,d\mathcal{L}^n
\end{aligned}$$

$$= \limsup_{m\to\infty} \frac{\beta_m \|\varphi\|_\infty}{\delta_m} \int_{\partial\Omega\cap C} |g^t - g^s| \, d\mathcal{H}^{n-1}$$

$$= \tilde{c}\|\varphi\|_\infty \int_{\partial\Omega\cap C} |g^t - g^s| \, d\mathcal{H}^{n-1} .$$

Since $\varepsilon > 0$ is arbitrary, the estimate is also true without ε. By (4.126), the right hand side tends to zero if $t, s \to 0$. Hence φg is ν^{int}-integrable and we have (4.130) (cf. [4, p. 114]). □

Lemma 4.53 *Let $\Omega \subset \mathbb{R}^n$ be an open bounded set with Lipschitz boundary, let $f \in \mathcal{BV}(\Omega)$ and let $f_k \in \mathcal{BV}(\Omega) \cap C^\infty(\Omega)$ be an approximating sequence satisfying* (4.119)*. Moreover let ν^{int} be the interior normal measure from Example 4.21. Then f is ν^{int}-integrable and*

$$\lim_{k\to\infty} ⨍_{\partial\Omega} f_k \varphi \, d\nu^{\text{int}} = ⨍_{\partial\Omega} f\varphi \, d\nu^{\text{int}} \tag{4.132}$$

for all $\varphi \in \mathcal{W}^{1,\infty}(\Omega, \mathbb{R}^n)$.

Proof We use the notation from the proof of Lemma 4.52 and, as there, it is sufficient to show (4.132) for the case where the integrals are restricted to an open cylinder $C := C(\bar{x}, r, h)$ for some $\bar{x} \in \partial\Omega$ and to work in the related coordinate system. Let us also fix some $\varphi \in \mathcal{W}^{1,\infty}(\Omega, \mathbb{R}^n)$.

(a) We start with some preliminaries. For $g \in \mathcal{BV}(\Omega)$ and $t, \tau > 0$ we set

$$g^x(t) := g^t(x) = g(x', \gamma(x') + t) , \quad g^{(t,\tau)}(x) := \frac{1}{\tau} \int_t^{t+\tau} g^s(x) \, ds .$$

Then there is some $\Gamma \subset \partial\Omega \cap C$ with $\mathcal{H}^{n-1}\big((\partial\Omega \cap C) \setminus \Gamma\big) = 0$ such that

$$g^x \in \mathcal{BV}\big((0, h)\big) \quad \text{for all } x \in \Gamma$$

(cf. [22, pp. 217, 220]). These g^x agree $\mathcal{L}^1$-a.e. with their right continuous representative. Thus we can identify g with a representative where

$$\text{all } g^x \text{ with } x \in \Gamma \text{ are continuous from the right}$$

(cf. [2, p. 136]). With the distributional derivative Dg^x, that is a Radon measure on $(0, h)$, we then have

$$g^x(s) = g^x(t) + Dg^x\big((t, s]\big) \quad \text{for all } t < s , \ x \in \Gamma$$

(cf. [2, pp. 136, 139]). Since $|Dg^x|\big((t,s]\big)$ is the total variation of g^x on $(t,s]$,

$$|g^x(s) - g^x(t)| \le |Dg^x|\big((t,s]\big) \quad \text{for all } \ x \in \Gamma\,. \tag{4.133}$$

The distributional derivative $D_n g$ with respect to x_n is a Radon measure on $\Omega \cap C$, since we have for all $0 \le t < s$

$$\begin{aligned} |D_n g|(C^{t,s}) &= \sup\Big\{ \int_{C^{t,s}} g \frac{\partial \psi}{\partial x_n}\, d\mathcal{L}^n \;\Big|\; \psi \in C_c^1(C^{t,s}),\ \|\psi\|_\infty \le 1 \Big\} \\ &\le \sup\Big\{ \int_{C^{t,s}} g \operatorname{div}\varphi\, d\mathcal{L}^n \;\Big|\; \varphi \in C_c^1(C^{t,s},\mathbb{R}^n),\ \|\varphi\|_\infty \le 1 \Big\} \\ &= |Dg|(C^{t,s}) \end{aligned} \tag{4.134}$$

(cf. [2, pp. 194, 195] and take above $\varphi = (0,\dots,0,\psi)$ to see the inequality). Therefore

$$|D_n g|(C^{t,s}) = \int_{\partial\Omega\cap C} |Dg^x|\big((t,s)\big)\, d\mathcal{H}^{n-1} < \infty \ \text{ for all } \ 0 \le t < s \tag{4.135}$$

(cf. [2, p. 195], [22, p. 220]). Since $|Dg|$ is a Radon measure on the open set $\Omega \cap C$,

$$\lim_{t\to 0} |Dg|(\overline{C^{0,t}}) = 0\,. \tag{4.136}$$

Using (4.133)–(4.135), we get for $t, \tau > 0$,

$$\begin{aligned} \int_{\partial\Omega\cap C} |g^t - g^{(t,\tau)}|\, d\mathcal{H}^{n-1} &= \int_{\partial\Omega\cap C} \Big| \frac{1}{\tau} \int_t^{t+\tau} g^t - g^s\, ds \Big|\, d\mathcal{H}^{n-1} \\ &\le \frac{1}{\tau} \int_t^{t+\tau} \int_{\partial\Omega\cap C} |g^t - g^s|\, d\mathcal{H}^{n-1}\, ds \\ &\le \frac{1}{\tau} \int_t^{t+\tau} \int_{\partial\Omega\cap C} |Dg^x|\big((t, s+\tau)\big)\, d\mathcal{H}^{n-1}\, ds \\ &= \frac{1}{\tau} \int_t^{t+\tau} |D_n g|(C^{t,s+\tau})\, ds \\ &\le |D_n g|(C^{t,t+2\tau}) \ \le\ |Dg|(C^{0,t+2\tau})\,. \end{aligned} \tag{4.137}$$

(b) We fix some $\tilde{\varepsilon} > 0$ and show that there is some $t_0 > 0$ and some $k_0 \in \mathbb{N}$ such that

$$\int_{\partial\Omega\cap C} |f_k^t - f^t|\, d\mathcal{H}^{n-1} \le \tilde{\varepsilon} \quad \text{for all } \ k > k_0\,,\ 0 < t < t_0\,. \tag{4.138}$$

By (4.136) we can choose some $t_0 > 0$ such that

$$|Df|(\overline{C^{0,3t_0}}) < \frac{\tilde{\varepsilon}}{4}.$$

Since

$$\limsup_{k\to\infty} |Df_k|(\overline{C^{0,3t_0}}) \le |Df|(\overline{C^{0,3t_0}})$$

(cf. [32, p. 93]), there is some $k_0 \in \mathbb{N}$ with

$$|Df_k|(\overline{C^{0,3t_0}}) \le |Df|(\overline{C^{0,3t_0}}) + \frac{\tilde{\varepsilon}}{4} \quad \text{for all } k > k_0 .$$

Let us fix some $\tau < t_0$. Then, by $f_k \to f$ in $\mathcal{L}^1(\Omega)$, we can assume that k_0 is so large that

$$\frac{1}{\tau}\int_{C^{0,2t_0}} |f_k - f|\, d\mathcal{L}^n < \frac{\tilde{\varepsilon}}{4} \quad \text{for all } k > k_0 .$$

Consequently, using (4.137), we obtain for $0 < t < t_0$ and $k > k_0$

$$\begin{aligned}
&\int_{\partial\Omega\cap C} |f_k^t - f^t|\, d\mathcal{H}^{n-1} \\
&\quad \le \int_{\partial\Omega\cap C} |f_k^t - f_k^{(t,\tau)}|\, d\mathcal{H}^{n-1} + \int_{\partial\Omega\cap C} |f_k^{(t,\tau)} - f^{(t,\tau)}|\, d\mathcal{H}^{n-1} \\
&\qquad + \int_{\partial\Omega\cap C} |f^{(t,\tau)} - f^t|\, d\mathcal{H}^{n-1} \\
&\quad \le |Df_k|(C^{0,t+2\tau}) + |Df|(C^{0,t+2\tau}) \\
&\qquad + \int_{\partial\Omega\cap C} \left| \frac{1}{\tau}\int_t^{t+\tau} f_k^s - f^s\, ds \right| d\mathcal{H}^{n-1} \\
&\quad \le |Df_k|(\overline{C^{0,3t_0}}) + |Df|(\overline{C^{0,3t_0}}) + \frac{1}{\tau}\int_{C^{t,t+\tau}} |f_k - f|\, d\mathcal{L}^n \\
&\quad \le 2|Df|(\overline{C^{0,3t_0}}) + \frac{\tilde{\varepsilon}}{4} + \frac{1}{\tau}\int_{C^{0,2t_0}} |f_k - f|\, d\mathcal{L}^n \le \tilde{\varepsilon}
\end{aligned}$$

which verifies (4.138).

(c) We show that $\varphi f_k \overset{\nu^{\text{int}}}{\to} \varphi f$. For $\varepsilon > 0$ we define

$$B_k := \{x \in \Omega \cap C \mid |f_k - f| > \varepsilon\},$$

$$B_k^t := \{x \in \partial\Omega \cap C \mid |f_k^t - f^t| > \varepsilon\}.$$

Let us also fix some $\tilde{\varepsilon} > 0$ and let $t_0 > 0$ and $k_0 \in \mathbb{N}$ be related to $\tilde{\varepsilon}$ according to (4.138). The Chebyshev inequality and (4.138) imply

$$\mathcal{H}^{n-1}(B_k^t) \le \frac{1}{\varepsilon} \int_{B_k^t} |f_k^t - f^t| \, d\mathcal{H}^{n-1} \le \frac{\tilde{\varepsilon}}{\varepsilon} \quad \text{for all } k > k_0\,, \ 0 < t < t_0\,.$$

With (4.31), (4.123), and $\beta_m \to 0$, we get

$$\begin{aligned}
|\nu^{\text{int}}|(B_k) &\le \limsup_{m\to\infty} \|D\chi_m\|_{\mathcal{L}^1(B_k)} = \limsup_{m\to\infty} \int_{B_k} |D\chi_m| \, d\mathcal{L}^n \\
&\le \limsup_{m\to\infty} \tfrac{1}{\delta_m} \int_{B_k \cap C^{0,\beta_m}} d\mathcal{L}^n \\
&\le \limsup_{m\to\infty} \tfrac{1}{\delta_m} \int_0^{\beta_m} \mathcal{H}^{n-1}(B_k^t)\, dt \\
&\le \limsup_{m\to\infty} \tfrac{\tilde{\varepsilon}\beta_m}{\varepsilon\delta_m} = \frac{\tilde{c}\tilde{\varepsilon}}{\varepsilon} \quad \text{for all } k > k_0\,.
\end{aligned}$$

Therefore $|\nu^{\text{int}}|(B_k) \to 0$ for all $\varepsilon > 0$ and hence $f_k \overset{\nu^{\text{int}}}{\longrightarrow} f$. For $\varphi f_k \overset{\nu^{\text{int}}}{\longrightarrow} \varphi f$ we argue as in part (e) of the proof of Lemma 4.52.

(d) Let us fix $\varepsilon > 0$ and let $t_0 > 0$, $k_0 \in \mathbb{N}$ be related to $\tilde{\varepsilon} > 0$ as in (4.138). By (2.9) there is some $\tilde{\varphi} \in \mathcal{L}^\infty(\Omega, \mathbb{R}^n)$ with $\|\tilde{\varphi}\|_\infty \le 1$ and by Theorem 4.19 there is a subsequence $\{\chi_{m'}\}$ such that for $k, l > k_0$ and $t < t_0$

$$\begin{aligned}
&\int_{\Omega\cap C} |f_k^t - f_l^t| \, d|\nu^{\text{int}}| - \varepsilon \\
&\quad \le \int_{\Omega\cap C} |f_k^t - f_l^t| \, \tilde{\varphi} \, d\nu^{\text{int}} \\
&\quad = \lim_{m'\to\infty} \int_{\Omega\cap C} |f_k^t - f_l^t| \, \tilde{\varphi} \cdot D\chi_{m'} \, d\mathcal{L}^n \\
&\quad \le \limsup_{m\to\infty} \frac{1}{\delta_m} \int_{C^{0,\beta_m}} |f_k^t - f_l^t| \, d\mathcal{L}^n \\
&\quad = \limsup_{m\to\infty} \frac{\beta_m}{\delta_m} \int_{\partial\Omega\cap C} |f_k^t - f_l^t| \, d\mathcal{H}^{n-1} \\
&\quad \le \tilde{c} \limsup_{m\to\infty} \int_{\partial\Omega\cap C} \big(|f_k^t - f^t| + |f^t - f_l^t|\big) \, d\mathcal{H}^{n-1} \\
&\quad = 2\tilde{c}\tilde{\varepsilon}\,.
\end{aligned}$$

This is true without $\varepsilon > 0$, since it is arbitrary.

(e) Again let $t_0 > 0$, $k_0 \in \mathbb{N}$ be related to $\tilde{\varepsilon} > 0$ as in (4.138). We combine the arguments from (f) in the proof of Lemma 4.52 with (4.124) and an estimate of the total variation from (b) in this proof to get, for $0 < s, t < t_0$,

$$\int_{\Omega\cap C} |f_k^t - f_k^s|\, d|\nu^{\text{int}}| \le \tilde{c} \int_{C^{0,t_0}} |Df_k|\, d\mathcal{L}^n \le \tilde{c}\tilde{\varepsilon} \quad \text{for all} \quad k > k_0.$$

Since $f_k^t \xrightarrow{\nu^{\text{int}}} f_k$ by the proof of Lemma 4.52, we can take $s \to 0$ to obtain

$$\int_{\Omega\cap C} |f_k^t - f_k|\, d|\nu^{\text{int}}| \le \tilde{c}\tilde{\varepsilon} \quad \text{for all} \quad k > k_0$$

(cf. also [4, p. 114]). Using (d) we obtain for some fixed $t \in (0, t_0)$

$$\begin{aligned}
&\int_{\Omega\cap C} |\varphi||f_k - f_l|\, d|\nu^{\text{int}}| \\
&\quad \le \|\varphi\|_\infty \int_{\Omega\cap C} |f_k - f_k^t|\, d|\nu^{\text{int}}| + \|\varphi\|_\infty \int_{\Omega\cap C} |f_k^t - f_l^t|\, d|\nu^{\text{int}}| \\
&\qquad + \|\varphi\|_\infty \int_{\Omega\cap C} |f_l^t - f_l|\, d|\nu^{\text{int}}| \\
&\quad \le 4\tilde{c}\tilde{\varepsilon}\|\varphi\|_\infty \quad \text{for all} \quad k, l > k_0\,.
\end{aligned}$$

Since $\tilde{\varepsilon} > 0$ is arbitrary, the left hand side tends to zero as $k, l \to \infty$. Using that $\varphi f_k \xrightarrow{\nu^{\text{int}}} \varphi f$ we obtain that φf is ν^{int}-integrable and the assertion (4.132) follows (cf. [4, p. 114]). □

As application of the introduced theory we finally consider a general boundary value problem for the p-Laplace operator. Let $\Omega \subset \mathbb{R}^n$ be an open bounded set and let $1 < p < \infty$. The trace operator $T : \mathcal{W}^{1,1}(\Omega) \to \mathcal{W}^{1,\infty}(\Omega)^*$ from Proposition 3.7 (with $U = \Omega$) is also a linear continuous operator on $\mathcal{W}^{1,p}(\Omega)$ by the continuous embedding $\mathcal{W}^{1,p}(U) \hookrightarrow \mathcal{W}^{1,1}(U)$. For given $g \in \mathcal{L}^{p'}(\Omega)$ and $f_b \in \mathcal{W}^{1,p}(\Omega)$ we call $f \in \mathcal{W}^{1,p}(\Omega)$ weak solution of the boundary value problem

$$-\operatorname{div}\big(|Df|^{p-2}Df\big) = g \quad \text{on } \Omega\,, \qquad f = f_b \quad \text{on } \partial\Omega \tag{4.139}$$

if we have that

$$\int_\Omega |Df|^{p-2}Df\,D\varphi - g\varphi\, d\mathcal{L}^n = 0 \quad \text{for all } \varphi \in C_c^\infty(\Omega) \quad \text{and}$$

$$\langle T(f - f_b), \varphi\rangle = 0 \quad \text{for all } \varphi \in \mathcal{W}^{1,\infty}(\Omega, \mathbb{R}^n)\,. \tag{4.140}$$

We show that this problem has always a solution without any regularity assumption on the boundary $\partial\Omega$. Before let us discuss the boundary condition (4.140) for Ω having Lipschitz boundary. From (4.114) we get

$$\langle T(f - f_b), \varphi\rangle = \operatorname{div}((f - f_b)\varphi)(\Omega) = \int_{\partial\Omega} (f^\times - f_b^\times)\varphi \cdot \nu \, d\mathcal{H}^{n-1} = 0$$

for all $\varphi \in \mathcal{W}^{1,\infty}(\Omega, \mathbb{R}^n)$ (identified with the continuous extension on $\partial\Omega$). By approximation, the most right equality is even valid for all $\varphi \in C(\overline{\Omega}, \mathbb{R}^n)$. Hence $(f^\times - f_b^\times)\nu\mathcal{H}^{n-1}\lfloor\partial\Omega$ has to be the zero measure. Consequently

$$f^\times = f_b^\times \quad \mathcal{H}^{n-1}\text{-a.e. on } \partial\Omega\,,$$

which is the usual pointwise boundary condition. Let us finally point out that the trace Tf is uniquely defined though its representation according to Theorem 4.40 is not.

Theorem 4.54 *Let $\Omega \subset \mathbb{R}^n$ be open and bounded, let $1 < p < \infty$, let $g \in \mathcal{L}^{p'}(\Omega)$ (where $\frac{1}{p} + \frac{1}{p'} = 1$), let $f_b \in \mathcal{W}^{1,p}(\Omega)$, and let T be the trace operator from Proposition 3.7. Then there is a weak solution $f \in \mathcal{W}^{1,p}(\Omega)$ of the boundary value problem* (4.139).

Notice that we obviously have that

$$Tf = 0 \quad \text{for all} \quad f \in C_c^\infty(\Omega)\,.$$

Thus, by the continuity of T on $\mathcal{W}^{1,p}(\Omega)$,

$$\mathcal{W}_0^{1,p}(\Omega) \subset \{f \in \mathcal{W}^{1,p}(\Omega) \mid Tf = 0\}\,. \tag{4.141}$$

It turns out to be sufficient for the theorem to study a variational problem on the set $f_b + \mathcal{W}_0^{1,p}(\Omega)$.

Proof of Theorem 4.54 We consider the minimization problem

$$E(f) := \int_\Omega |Df|^p - fg \, d\mathcal{L}^n \;\to\; \text{Min!}\,, \quad f \in \mathcal{W}^{1,p}(\Omega)$$

subject to

$$M := \left\{f \in \mathcal{W}^{1,p}(\Omega) \,\middle|\, f = f_b + f_0\,,\; f_0 \in \mathcal{W}_0^{1,p}(\Omega)\right\}.$$

Let $f_k \in \mathcal{W}^{1,p}(\Omega)$ be a minimizing sequence $f_k \in \mathcal{W}^{1,p}(\Omega)$. Then, by the Poincaré inequality, there is some $c > 0$ such that

$$\begin{aligned}\|f_k\|_{\mathcal{L}^p} &\leq \|f_k - f_b\|_{\mathcal{L}^p} + \|f_b\|_{\mathcal{L}^p} \\ &\leq c\|Df_k - Df_b\|_{\mathcal{L}^p} + \|f_b\|_{\mathcal{L}^p} \\ &\leq c\|Df_k\|_{\mathcal{L}^p} + c\|Df_b\|_{\mathcal{L}^p} + \|f_b\|_{\mathcal{L}^p}\,.\end{aligned}$$

Consequently, for some $\tilde{c} > 0$,

$$\begin{aligned}E(f_k) &\geq \|Df_k\|^p_{\mathcal{L}^p} - \|g\|_{\mathcal{L}^{p'}}\|f_k\|_{\mathcal{L}^p} \\ &\geq \|Df_k\|^p_{\mathcal{L}^p} - \tilde{c}\big(\|Df_k\|_{\mathcal{L}^p} + \|Df_b\|_{\mathcal{L}^p} + \|f_b\|_{\mathcal{L}^p}\big) \\ &= \|Df_k\|_{\mathcal{L}^p}\big(\|Df_k\|^{p-1}_{\mathcal{L}^p} - \tilde{c}\big) - \tilde{c}\big(\|Df_b\|_{\mathcal{L}^p} + \|f_b\|_{\mathcal{L}^p}\big)\,.\end{aligned}$$

Combining both estimates we get that the f_k must be bounded in $\mathcal{W}^{1,p}(\Omega)$. Thus there is a weakly convergent subsequence, denoted the same way, with $f_k \rightharpoonup: f$. Since M is a closed affine subspace of $\mathcal{W}^{1,p}(\Omega)$, it is also weakly closed. Therefore $f \in M$. As convex and continuous function, E is weakly lower semicontinuous (cf. [18, p. 49 or 74]). This implies that f solves the minimization problem. Obviously, $f = f_b + f_0$ for some $f_0 \in \mathcal{W}^{1,p}_0(\Omega)$. Hence

$$\langle T(f - f_b), \varphi\rangle = \langle Tf_0, \varphi\rangle = 0 \quad \text{for all } \varphi \in \mathcal{W}^{1,\infty}(\Omega, \mathbb{R}^n)$$

and thus, f satisfies the boundary condition.

Now we decompose $E = E_1 - E_2$ in the obvious way where E_1 is convex and E_2 is linear and continuous. Clearly,

$$E_2'(f, \varphi) = \int_\Omega g\varphi\, d\mathcal{L}^n \quad \text{for all } \varphi \in \mathcal{W}^{1,p}_0(\Omega)\,.$$

Moreover E_1 is Gâteaux differentiable on M with

$$E_1'(f, \varphi) = \int_\Omega |Df|^{p-2} Df\, D\varphi\, d\mathcal{L}^n \quad \text{for all } \varphi \in \mathcal{W}^{1,p}_0(\Omega)$$

(cf. [18, p. 89]). Since f minimizes E on M,

$$E'(f, \varphi) = \int_\Omega |Df|^{p-2} Df\, D\varphi - g\varphi\, d\mathcal{L}^n = 0 \quad \text{for all } \varphi \in \mathcal{W}^{1,p}_0(\Omega)\,.$$

Consequently, f is a weak solution of (4.139) and the proof is complete. □

References

1. Adams, R.A.: Sobolev Spaces. Academic Press, Boston (1978)
2. Ambrosio, L., Fusco, N., Pallara, D.: Functions of Bounded Variation and Free Discontinuity Problems. Oxford University Press, New York (2000)
3. Anzellotti, G.: Pairings between measures and bounded functions and compensated compactness. Ann. Mat. Pura Appl. (4) **135**, 193–318 (1983)
4. Bhaskara Rao, K.P.S., Bhaskara Rao, M.: Theory of Charges - A Study of Finitely Additive Measures. Pure and Applied Mathematics - A Series of Monographs and Textbooks. Academic Press, London (1983)
5. Chen, G.-Q., Frid, H.: Divergence-measure fields and hyperbolic conservation laws. Arch. Ration. Mech. Anal. **147**, 89–118 (1999)
6. Chen, G.-Q., Frid, H.: On the theory of divergence-measure fields and its applications. Bull. Braz. Math. Soc. (N.S.) **32**, 401–433 (2001)
7. Chen, G.-Q, Frid, H.: Extended divergence-measure fields and the Euler equations for gas dynamics. Commun. Math. Phys. **236**, 251–280 (2003)
8. Chen, G.-Q., Torres, M.: Divergence-measure fields, sets of finite perimeter, and conservation laws. Arch. Ration. Mech. Anal. **175**, 245–267 (2005)
9. Chen, G.-Q., Torres, M.: On the structure of solutions of nonlinear hyperbolic systems of conservation laws. Commun. Pure Appl. Anal. **10**, 1011–1036 (2011)
10. Chen, G.-Q., Torres, M., Ziemer, W.P.: Gauss-Green theorem for weakly differentiable vector fields, sets of finite perimeter, and balance laws. Comm. Pure Appl. Math. **62**, 242–304 (2009)
11. Chen, G.-Q., Comi, G., Torres, M.: Cauchy fluxes and Gauss-Green formulas for divergence-measure fields over general open sets. Arch. Ration. Mech. Anal. **233**, 87–166 (2019)
12. Chen, G.-Q., Li, Q., Torres, M.: Traces and extensions of bounded divergence-measure fields on rough open sets. Indiana Univ. Math. J. **69**, 229–264 (2020)
13. Comi, G.E., Payne, K.R.: On locally essentially bounded divergence measure fields and sets of locally finite perimeter. Adv. Calc. Var. **13**, 179–217 (2020)
14. Comi, G.E., Torres, M.: One-sided approximation of sets of finite perimeter. Rend. Lincei Mat. Appl. **28**, 181–190 (2017)
15. Crasta, G., De Cicco, V.: Anzelotti's pairing theory and the Gauss-Green theorem. Adv. Math. **343**, 935–970 (2019)
16. Crasta, G., De Cicco, V.: An extension of the pairing theory between divergence-measure fields and BV functions. J. Funct. Anal. **276**, 2605–2635 (2019)
17. Crasta, G., De Cicco, V., Malusa, A.: Pairings between bounded divergence-measure vector fields and BV functions. Adv. Calc. Var. **15**, 787–810 (2022)
18. Dacorogna, B.: Direct Methods in the Calculus of Variations. Springer, Berlin (1989)

F. Schuricht, M. Schönherr, *A Theory of Traces and the Divergence Theorem*,
Lecture Notes in Mathematics 2372, https://doi.org/10.1007/978-3-031-86664-7

19. De Giorgi, E.: Nuovi teoremi relativi alle misure $(r-1)$-dimensionali in uno spazio ad r dimensioni. Ric. Mat. **4**, 95–113 (1955)
20. Degiovanni, M., Marzocchi, A., Musesti, A.: Cauchy fluxes associated with tensor fields having divergence measure. Arch. Ration. Mech. Anal. **147**, 197–223 (1999)
21. Dunford, N., Schwartz, J.T.: Linear Operators. Part I: General Theory. Wiley Interscience, New York (1988)
22. Evans, L.C., Gariepy, R.: Measure Theory and Fine Properties of Functions. CRC Press, Boca Raton (1992)
23. Federer, H.: The Gauss-Green theorem. Trans. Am. Math. Soc. **58**, 44–76 (1945)
24. Federer, H.: A note on the Gauss-Green theorem. Proc. Amer. Math. Soc. **9**, 447–451 (1958)
25. Federer, H.: Geometric Measure Theory. Springer, New York (1969)
26. Infeld, L.: Quest. An Autobiography, 2nd edn., p. 279. Chelsea Publishing Company, New York (1980)
27. Kawohl, B., Schuricht, F.: Dirichlet problems for the 1-Laplace operator, including the eigenvalue problem. Commun. Contemp. Math. **9**, 515–543 (2007)
28. Kolmogoroff, A.: Untersuchungen über den Integralbegriff. Math. Ann. **103**, 654–696 (1930)
29. Kraft, D.: Measure-theoretic properties of level sets of distance functions. J. Geom. Anal. **26**, 2777–2796 (2016)
30. Leonardi, G.P., Saracco, G.: Rigidity and trace properties of divergence-measure vector fields. Adv. Calc. Var. **15**, 133–149 (2022)
31. Maggi, F.: Sets of Finite Perimeter and Geometric Variational Problems. Cambridge University Press, Cambridge (2012)
32. Pfeffer, W.F.: The Divergence Theorem and Sets of Finite Perimeter. CRC Press, Boca Raton (2012)
33. Podio-Guidugli, P.: Examples of concentrated contact interactions in simple bodies. J. Elasticity **75**, 167–186 (2004)
34. Podio-Guidugli, P., Schuricht, F.: Concentrated actions on cuspidate plane bodies. J. Elasticity **106**, 107–114 (2011)
35. Schönherr, M.: Pure measures, traces and a general theorem of Gauss. Doctoral thesis, TU Dresden (2017)
36. Schönherr, M., Schuricht, F.: A general theorem of Gauss using pure measures. Preprint. arXiv:1710.02211 (2017)
37. Schönherr, M., Schuricht, F.: Pure measures, density measures and the dual of L^∞. Preprint. arXiv:1710.02197 (2017)
38. Schuricht, F.: A new mathematical foundation for contact interactions in continuum physics. Arch. Ration. Mech. Anal. **184**, 495–551 (2007)
39. Schuricht, F.: Interactions in continuum physics. In: Šilhavý, M. (ed.) Mathematical Modelling of Bodies with Complicated Bulk and Boundary Behavior. Quad. Mat. **20**, 169–196 (2007)
40. Šilhavý, M.: Cauchy's stress theorem and tensor fields with divergence measure in $\mathcal{L}^p$. Arch. Ration. Mech. Anal. **116**, 223–255 (1991)
41. Šilhavý, M.: Divergence measure fields and Cauchy's stress theorem. Rend. Sem. Mat. Padova **113**, 15–45 (2005)
42. Šilhavý, M.: Cauchy's stress theorem for stresses represented by measures. Continuum Mech. Thermodyn. **20**, 75–96 (2008)
43. Šilhavý, M.: The divergence theorem for divergence measure vectorfields on sets with fractal boundaries. Math. Mech. Solids **14**, 445–455 (2009)
44. Šilhavý, M.: The Gauss-Green theorem for bounded vectorfields with divergence measure on sets of finite perimeter. Czech Academy of Sciences, Institute of Mathematics, Preprint No. 27–2019, Praha (2019)
45. Stolze, C.H.: A history of the divergence theorem. Historia Math. **5**, 437–442 (1978)
46. Toland, J.: The Dual of $L_\infty(X, \mathcal{L}, \lambda)$, Finitely Additive Measures and Weak Convergence. A Primer. Springer Nature, Cham (2020)
47. Vol'pert, A.I.: The spaces BV and quasilinear equations. Mat. Sb. (N.S.) **73** (115), 255–302 (1967)

48. Vol'pert, A.I., Hudjaev, S.I.: Analysis in Classes of Discontinuous Functions and Equations of Mathematical Physics. Martinus Nijhoff Publishers, Dordrecht (1985)
49. Werner, D.: Funktionalanalysis. Springer, Berlin (2000)
50. Yosida, K.: Functional Analysis. Springer, Berlin (1980)
51. Yosida, K., Hewitt, E.: Finitely additive measures. Trans. Amer. Math. Soc. **72**, 46–66 (1952)
52. Zeidler, E.: Nonlinear Functional Analysis and Its Applications I. Fixed-Point Theorems. Springer, New York (1986)
53. Zeidler, E.: Nonlinear Functional Analysis and Its Applications II/B. Nonlinear Monotone Operators. Springer, New York (1990)
54. Zeidler, E.: Applied Functional Analysis. Applications to Mathematical Physics. Springer, New York (1995)
55. Zeidler, E.: Applied Functional Analysis. Main Principles and Their Applications. Springer, New York (1995)

Notation

! Warning

Notice that some of our terminology deviates from standard usage in measure theory (cf. Sect. 2.1). In particular the notion *measure* is used for any finitely additive measure while σ*-measure* is used for σ-additive measures.

Sets and vectors

$\mathbb{R}$, $\overline{\mathbb{R}}$, $\mathbb{R}_{\geq 0}$, $\overline{\mathbb{R}}_{\geq 0}$	$(-\infty, +\infty)$, $[-\infty, +\infty]$, $[0, \infty)$, $[0, \infty]$
$\widehat{\mathbb{N}}$	$\mathbb{N} \cup \{\infty\}$, 16
$\widehat{\mathbb{R}^n}$	$\mathbb{R}^n \cup \{\infty\}$, 16
$a \cdot b$, $A : B$	scalar product of vectors a, b and matrices A, B
$\lvert a\rvert_p$, $\lvert a\rvert$	p-norm, Euclidean norm for $a \in \mathbb{R}^n$
A^c	complement of set A
$\operatorname{int}\Omega$, $\operatorname{ext}\Omega$	topological interior, exterior of $\Omega \subset \mathbb{R}^n$
$\operatorname{int}_*\Omega$, $\operatorname{ext}_*\Omega$	measure theoretic interior, exterior of $\Omega \subset \mathbb{R}^n$
$\partial\Omega$, $\overline{\Omega}$	topological boundary, closure of $\Omega \subset \mathbb{R}^n$
$\partial_*\Omega$, $\partial^*\Omega$	measure theoretic, reduced boundary of $\Omega \subset \mathbb{R}^n$
$\operatorname{Per}(\Omega)$	perimeter of $\Omega \subset \mathbb{R}^n$
$\operatorname{dist}_\Omega(x)$	distance of x from Ω (equals $\inf_{y\in\Omega}\lvert x-y\rvert$)
Ω_δ	open δ-neighborhood of Ω for $\delta \in \mathbb{R}$ (equals $\{x \mid \operatorname{dist}_\Omega(x) - \operatorname{dist}_{\Omega^c}(x) < \delta\}$)
$B_r(x)$	open ball of radius r centered at x
$B_r^A(x)$	$B_r(x) \cap A$
$\Omega \Subset U$	Ω compactly contained in U, 10
$\mathcal{A}$, $\mathcal{A}^\sigma$	algebra, σ-algebra of sets

F. Schuricht, M. Schönherr, *A Theory of Traces and the Divergence Theorem*, Lecture Notes in Mathematics 2372, https://doi.org/10.1007/978-3-031-86664-7

$\mathcal{B}(U), \mathcal{P}(U)$	Borel subsets, power set of U
Γ	Borel set $\Gamma \subset \overline{\Omega}$, 25

Functions

$\operatorname{supp} f$	support of f
$f_{\lvert A}$	restriction of f to A
$(f)_{x,r}$	$\fint_{B_r(x)} f \, d\mathcal{L}^n$
Df	derivative of f
$\operatorname{Lip}(f)$	Lipschitz constant of f, 77
$\operatorname{div} F$	(distributional) divergence of vector field F, 10
$f_k \xrightarrow{\mu} f$	f_k converge in measure to f, 17
$f = g$ i.m.	f and g agree in measure, 17
$f \leq g$ i.m.	$f \leq g$ in measure, 17
$f = g$ a.e.	f and g agree almost everywhere, 17
$f \leq g$ a.e.	$f \leq g$ almost everywhere, 17
$\int_\Omega f \, d\mu$	integral of a scalar function, 18
$\int_\Omega F \, d\mu$	integral of a vector field, 19
$\fint_C f \, d\mu, \fint_C F \, d\mu$	integral is to take on any open neighborhood of C, 19
$\fint_M f \, d\mu$	$\frac{1}{\lvert\mu(M)\rvert} \int_M f \, d\mu \quad (= 0 \text{ if } \mu(M) = 0)$
$\operatorname{ess\,lim}_{\delta \downarrow 0}$	essential limit, 28
$\operatorname{ap\,lim}_{y \to x} f(y)$	approximate limit of f at x, 28
$\operatorname{ap\,lim\,inf}_{y \to x} f(y)$	lower approximate limit of f at x, 30
$\operatorname{ap\,lim\,sup}_{y \to x} f(y)$	upper approximate limit of f at x, 30
$f^{\pm}$	positive/negative part of function f
$f^i(x), f^s(x)$	lower and upper approximate limit of f at x, 30
$f^{\times}$	precise representative of f, 29
$\varphi_{\wr\Gamma}$	equivalence class of $\varphi \in \mathcal{L}^\infty$ in $\mathcal{L}^\infty_\Gamma$, 45
η_ε	standard mollifier on $B_\varepsilon(0)$, 106
χ_A	characteristic function of set A
χ_δ^Γ	tent function for Γ and $\delta > 0$, 54
χ	a good approximation of χ_Ω, 99
ν^Ω	unit normal field of Ω or unit normal on $\partial^*\Omega$, 70
$f^{\text{int}}, f^{\text{ext}}$	normal trace functions for $F \in \mathcal{DM}^\infty(U)$, 126
$F^{\text{int}}, F^{\text{ext}}$	normal trace functions for $f \in \mathcal{BV}(U)$, 146
T	trace operator on $\mathcal{DM}^1(U)$, 45

Spaces and norms

X^*	dual of normed space X
$\operatorname{ba}(\Omega, \mathcal{A})$	bounded measures, 12
$\operatorname{pa}(\Omega, \mathcal{A})$	pure measures, 12
$\operatorname{ca}(\Omega, \mathcal{A})$	countably additive (i.e. σ-additive) measures, 12

$\mathrm{ba}(\Omega, \mathcal{A}, \lambda)$	$\mu \in \mathrm{ba}(\Omega, \mathcal{A})$ with $\mu \ll^{w} \lambda$, 15
$C(\Omega)$	continuous functions on Ω
$C(\overline{\Omega})$	$f \in C(\Omega)$ with continuous extension on $\overline{\Omega}$
$C_c(\Omega)$	$f \in C(\Omega)$ with compact support
$C^k(\Omega)$	k-times continuously differentiable functions on Ω
$C^k(\overline{\Omega})$	$f \in C^k(\Omega)$ and all derivatives (order $\leq k$) belong to $C(\overline{\Omega})$
$C_c^k(\Omega)$	$f \in C^k(\Omega)$ with compact support
$C_c^\infty(\Omega)$	$f \in C_c^k(\Omega)$ for all $k \in \mathbb{N}$
$C(\Omega, \mathbb{R}^m)$ etc.	vector-valued functions mapping into $\mathbb{R}^m$
$\mathrm{Lip}(\Omega)$	Lipschitz continuous functions on Ω, 10
$L^p(\Omega, \mathcal{A}, \mu)$	p-integrable functions w.r.t. μ, 19
$L^\infty(\Omega, \mathcal{A}, \mu)$	essentially bounded functions w.r.t. μ, 20
$L^p(\Omega, \mu)$	$L^p(\Omega, \mathcal{B}(\Omega), \mu)$, 20
$L^p(\Omega)$	$L^p(\Omega, \mathcal{B}(\Omega), \mathcal{L}^n)$, 20
$L^p(\Omega, \mathcal{B}(\Omega), \mu)^m$	vector-valued version of $L^p(\Omega, \mathcal{B}(\Omega), \mu)$, 20
$L^p(\Omega, \mathbb{R}^m)$ etc.	vector-valued functions $L^p(\Omega)^m$, 20
$\mathcal{L}^p(\Omega, \mathcal{A}, \mu)$	equivalence classes i.m. of $L^p(\Omega, \mathcal{A}, \mu)$, 20
$\mathscr{L}^p(\Omega, \mathcal{A}, \mu)$	completion of $\mathcal{L}^p(\Omega, \mathcal{A}, \mu)$, 20
$\mathcal{L}_\Gamma^\infty(U, \mathbb{R}^m)$	factor space of $\mathcal{L}^\infty(U, \mathbb{R}^m)$ w.r.t. $\|\varphi\|_\Gamma = 0$, 45
$W^{k,p}(\Omega)$	Sobolev functions with weak derivatives in L^p
$W^{k,p}(\Omega, \mathbb{R}^m)$ etc.	vector-valued functions $W^{k,p}(\Omega)^m$
$\mathcal{W}^{k,p}(\Omega)$	equivalence classes of $W^{k,p}(\Omega)$
$\mathcal{W}_0^{k,p}(\Omega)$	completion of $C_c^\infty(\Omega)$ within $\mathcal{W}^{k,p}(\Omega)$
$BV(\Omega)$	space of functions of bounded variation
$\mathcal{BV}(\Omega)$	equivalence classes of $BV(\Omega)$
$\mathcal{M}(\Omega)$	space of finite Radon measures on Ω
$\mathcal{DM}^p(U)$	$\mathcal{L}^p$-vector fields where the divergence is a measure, 45
$\|\cdot\|_X$	norm on space X
$\|\cdot\|$	standard norm according to the context
$\|f\|_p$	L^p-norm for scalar functions, 19
$\|\lambda\|_p$	norm on completion $\mathscr{L}^P$, 20
$\|F\|_p$	L^p-norm for vector fields, 44
$\|F\|_{\mathcal{W}^{1,\infty}}$	norm on $\mathcal{W}^{1,\infty}(U, \mathbb{R}^m)$, 44
$\|F\|_{\mathcal{W}^{1,p}}$	norm on $\mathcal{W}^{1,p}(U, \mathbb{R}^m)$, 44
$\|f\|_{\mathcal{BV}}$	norm on $\mathcal{BV}(U)$, 44
$\|\varphi\|_{\mathrm{Lip}(\Omega)}$	norm on $\mathrm{Lip}(\Omega)$, 77
$\|\varphi\|_\Gamma$	semi norm on $\mathcal{L}^\infty$, 25
$\|\varphi_\Gamma\|$	norm on $\mathcal{L}_\Gamma^\infty$ $(= \|\varphi\|_\Gamma)$, 52
$\|F\|_{\mathcal{DM}^p}$	norm on $\mathcal{DM}^p$, 45
$\|\mu\|$	norm on $\mathrm{ba}(\Omega, \mathcal{A})^m$, 44
p'	Hölder conjugate of p

Measures

$\mathcal{L}^n$	Lebesgues measure on $\mathbb{R}^n$
$\mathcal{H}^k$	k-dimensional Hausdorff measure on $\mathbb{R}^n$
μ, λ	general (finitely additive) measures, 11
σ	σ-additive measure, 11
$\mu^{\pm}$	positive/negative part of μ, 12
$\lvert\mu\rvert$	total variation of μ, 12
μ^*	outer measure of μ, 12
$\mu\lfloor A$	restriction of μ to set A
$(f\mu)(A)$	$\int_A f\,d\mu$, 20
$\mu \perp \lambda$	μ and λ are singular, 13
$\mu \perp^s \lambda$	μ and λ are strongly singular, 13
$\mu \ll \lambda$	μ is absolutely continuous with respect to λ, 15
$\mu \ll^w \lambda$	μ is weakly absolutely continuous with respect to λ, 15
μ_c, μ_p	σ-additive, pure part of $\mu \in \mathrm{ba}(\Omega, \mathcal{A})$, 13
core μ	core of μ, 16
supp σ	support of σ-measure σ
dens_x	density measure at x, 14
dens_x^E	density measure at x w.r.t. set E, 26
δ_x	Dirac measure at x, 25
(λ_f, μ_F)	normal trace of F on $\partial\Omega$, 69
ν	normal measure, 100
$\nu^{\mathrm{int}}, \nu^{\mathrm{intc}}, \tilde{\nu}^{\mathrm{int}}$	interior normal measures, 108
$\nu^{\mathrm{ext}}, \nu^{\mathrm{extc}}, \tilde{\nu}^{\mathrm{ext}}$	exterior normal measures, 110
$\nu^{\mathrm{sym}}, \tilde{\nu}^{\mathrm{sym}}$	symmetric normal measure, 111
$\omega_{\partial\Omega}^{\mathrm{int}}$ etc.	area density for certain normal measures, 119

Others

$\iota(\varphi)$	$\iota : \mathcal{W}^{1,\infty} \to X_0 \times (\mathcal{L}^\infty)^n$ with $\iota(\varphi) = (\iota_0(\varphi), D\varphi)$, 53
$\iota_0(\varphi)$	$\iota_0 : \mathcal{W}^{1,\infty} \to X_0$, 53
case (G)	$X_0 = \mathcal{L}^\infty(\Gamma_\delta \cap U, \mathbb{R}^m)$ and $\iota_0(\varphi) = \varphi$, 55
case (L)	$X_0 = \mathcal{L}^\infty_\Gamma(\Gamma_\delta \cap U, \mathbb{R}^m)$ and $\iota_0(\varphi) = \varphi_{\wr\Gamma}$, 55
case (C)	$X_0 = C(\Gamma, \mathbb{R}^m)$, $\iota_0(\varphi) = \varphi_{\mid\Gamma}$, φ cont. extendable to Γ, 55

Index

F. Schuricht, M. Schönherr, *A Theory of Traces and the Divergence Theorem*,
Lecture Notes in Mathematics 2372, https://doi.org/10.1007/978-3-031-86664-7

LECTURE NOTES IN MATHEMATICS Springer

Editors in Chief: J.-M. Morel, B. Teissier;

Editorial Policy

1. Lecture Notes aim to report new developments in all areas of mathematics and their applications – quickly, informally and at a high level. Mathematical texts analysing new developments in modelling and numerical simulation are welcome.

   Manuscripts should be reasonably self-contained and rounded off. Thus they may, and often will, present not only results of the author but also related work by other people. They may be based on specialised lecture courses. Furthermore, the manuscripts should provide sufficient motivation, examples and applications. This clearly distinguishes Lecture Notes from journal articles or technical reports which normally are very concise. Articles intended for a journal but too long to be accepted by most journals, usually do not have this "lecture notes" character. For similar reasons it is unusual for doctoral theses to be accepted for the Lecture Notes series, though habilitation theses may be appropriate.

2. Besides monographs, multi-author manuscripts resulting from SUMMER SCHOOLS or similar INTENSIVE COURSES are welcome, provided their objective was held to present an active mathematical topic to an audience at the beginning or intermediate graduate level (a list of participants should be provided).

   The resulting manuscript should not be just a collection of course notes, but should require advance planning and coordination among the main lecturers. The subject matter should dictate the structure of the book. This structure should be motivated and explained in a scientific introduction, and the notation, references, index and formulation of results should be, if possible, unified by the editors. Each contribution should have an abstract and an introduction referring to the other contributions. In other words, more preparatory work must go into a multi-authored volume than simply assembling a disparate collection of papers, communicated at the event.

3. Manuscripts should be submitted either online at www.editorialmanager.com/lnm to Springer's mathematics editorial in Heidelberg, or electronically to one of the series editors. Authors should be aware that incomplete or insufficiently close-to-final manuscripts almost always result in longer refereeing times and nevertheless unclear referees' recommendations, making further refereeing of a final draft necessary. The strict minimum amount of material that will be considered should include a detailed outline describing the planned contents of each chapter, a bibliography and several sample chapters. Parallel submission of a manuscript to another publisher while under consideration for LNM is not acceptable and can lead to rejection.

4. In general, **monographs** will be sent out to at least 2 external referees for evaluation.

   A final decision to publish can be made only on the basis of the complete manuscript, however a refereeing process leading to a preliminary decision can be based on a pre-final or incomplete manuscript.

   Volume Editors of **multi-author works** are expected to arrange for the refereeing, to the usual scientific standards, of the individual contributions. If the resulting reports can be forwarded to the LNM Editorial Board, this is very helpful. If no reports are forwarded or if other questions remain unclear in respect of homogeneity etc, the series editors may wish to consult external referees for an overall evaluation of the volume.

5. Manuscripts should in general be submitted in English. Final manuscripts should contain at least 100 pages of mathematical text and should always include

   - a table of contents;
   - an informative introduction, with adequate motivation and perhaps some historical remarks: it should be accessible to a reader not intimately familiar with the topic treated;
   - a subject index: as a rule this is genuinely helpful for the reader.
   - For evaluation purposes, manuscripts should be submitted as pdf files.

6. Careful preparation of the manuscripts will help keep production time short besides ensuring satisfactory appearance of the finished book in print and online. After acceptance of the manuscript authors will be asked to prepare the final LaTeX source files (see LaTeX templates online: https://www.springer.com/gb/authors-editors/book-authors-editors/manuscriptpreparation/5636) plus the corresponding pdf- or zipped ps-file. The LaTeX source files are essential for producing the full-text online version of the book, see http://link.springer.com/bookseries/304 for the existing online volumes of LNM). The technical production of a Lecture Notes volume takes approximately 12 weeks. Additional instructions, if necessary, are available on request from lnm@springer.com.

7. Authors receive a total of 30 free copies of their volume and free access to their book on SpringerLink, but no royalties. They are entitled to a discount of 33.3 % on the price of Springer books purchased for their personal use, if ordering directly from Springer.

8. Commitment to publish is made by a *Publishing Agreement*; contributing authors of multiauthor books are requested to sign a *Consent to Publish form*. Springer-Verlag registers the copyright for each volume. Authors are free to reuse material contained in their LNM volumes in later publications: a brief written (or e-mail) request for formal permission is sufficient.

Addresses:

Professor Jean-Michel Morel, CMLA, École Normale Supérieure de Cachan, France
E-mail: moreljeanmichel@gmail.com

Professor Bernard Teissier, Equipe Géométrie et Dynamique,
Institut de Mathématiques de Jussieu – Paris Rive Gauche, Paris, France
E-mail: bernard.teissier@imj-prg.fr

Springer: Ute McCrory, Mathematics, Heidelberg, Germany,
E-mail: lnm@springer.com

The manufacturer's authorised representative in the EU is Springer Nature Customer Service Centre GmbH, Europaplatz 3, 69115 Heidelberg, Germany. If you have any concerns regarding our products, please contact ProductSafety@springernature.com

Printed and bound by CPI Group (UK) Ltd, Croydon, CR0 4YY
26/03/2026
02078976-0005